Anaerobic infections

A SCOPE® PUBLICATION

Sydney M. Finegold, MD
Vera L. Sutter, PhD

From the Medical and Research Services,
Veterans Administration, Wadsworth Medical Center,
and the Department of Medicine,
UCLA Medical Center, Los Angeles, California.

Dosage schedules recommended are those of the authors; they may be changed as clinical experience accumulates.

Please check the package insert before prescribing an antibacterial drug because indications, recommended dosages, and possible side effects are subject to change.

Library of Congress Card Number: 72-79754 ISBN 0-89501-010-0

Fifth Edition, 1983

Contents

Preface

In preparing the fourth edition of this monograph, our primary purpose is again to present an overview of anaerobic infections. Anaerobic infections have been the most commonly overlooked of all bacterial infections because clinicians and microbiologists alike have not always been cognizant of the importance of anaerobic organisms. Recently, many groups have intensively examined anaerobes and the infections produced by them, and the resultant increase in publications has stimulated still further interest. Furthermore, simplified anaerobic techniques now make it feasible for even small clinical laboratories to isolate anaerobic bacteria and to characterize them with reasonable accuracy into broad groupings. As a result of all this interest and activity, the role of anaerobic bacteria in pathophysiologic processes is becoming better understood.

We are grateful to those who permitted us to use their illustrations. In particular, we thank Martin McHenry, MD, of the Cleveland Clinic; Fred Killefer, MD, formerly of the Wadsworth VA Hospital and UCLA Medical Center; and V. R. Dowell, Jr , PhD, of the Center for Disease Control, Atlanta, for their generous contributions of excellent photographs. We also thank our former infectious disease Fellows, particularly Francis Tally, MD, for their assistance in obtaining illustrations for this monograph, and we thank Diane M. Citron for her assistance in providing illustrations. We express our appreciation to the VA Wadsworth Medical Center's Medical Media Service for the excellent photographic work. Finally, we gratefully acknowledge the important contributions of Jon E. Rosenblatt, MD, and Howard R. Attebery, DDS, to earlier editions of this publication.

Sydney M. Finegold, MD
Vera L. Sutter, PhD

Introduction

Anaerobic infections are common and significant; essentially every type of bacterial infection that occurs in humans can involve anaerobic bacteria. In many types of infections, anaerobes are the main etiologic agents. Certain anaerobic infections are fulminating and most tend to produce tissue necrosis. The course of these infections is frequently prolonged, with significant attendant mortality. Occasionally residua of infection occur.

Because good treatment depends on early specific diagnosis, clinicians must be aware of the characteristics of anaerobic infections and of the proper methods for obtaining specimens and submitting them to the laboratory. Clinical microbiologists, likewise, must be able to culture, isolate, and correctly identify any anaerobes that are involved.

What is an anaerobe? A simplified working definition is that an anaerobe is a microorganism that requires a reduced oxygen tension for growth and cannot grow on solid media in an environment of 10% CO_2 in air (18% oxygen); some anaerobes require the complete or almost complete absence of molecular oxygen for survival. This definition, however, does not consider oxidation-reduction requirements, susceptibility to peroxides, and other conditions that may be very important for the survival of anaerobes.

Tolerance of anaerobic bacteria to oxygen and oxidation-reduction potentials varies. For example, anaerobes such as *Clostridium perfringens* are relatively tolerant both to oxygen and to somewhat high oxidation-reduction potentials. Consequently, they can grow in less-than-ideal anaerobic systems. Other species of *Clostridium* and some species of *Eubacterium* and other genera are much more fastidious anaerobes and require the absence of oxygen throughout the cultivation process, from preparation of the media to identification of the organism.

Unlike anaerobes, facultative organisms – such as *Escherichia coli* – can grow in either the presence or absence of air, and microaerophilic organisms – such as *Actinomyces naeslundii*, *Arachnia propionica*, and some gram-positive cocci – prefer a reduced oxygen tension but can grow on solid media in 10% CO_2 in air (some of these organisms may actually be capnophilic rather than microaerophilic).

As methods for identifying and classifying microorganisms continue to become more sophisticated, new organisms are discovered, and recently discovered or established organisms may be reclassified. Consequently, literature dealing with the taxonomy and nomenclature of many anaerobes is very confusing. For example, among the medically important gram-negative bacilli, organisms that were classified as *Sphaerophorus* for many years are now classified as *Fusobacterium* in the latest edition (eighth) of *Bergey's Manual of Determinative Bacteriology*. Among the medically important gram-positive bacilli, anaerobic organisms that were classified as *Corynebacterium* are now classified as *Propionibacterium*. Individual species such as *Actinomyces eriksonii*, *Catenabacterium catenaforme*, and *Catenabacterium filamentosum* are now classified in other genera. Appendix A lists the newer nomenclature for various anaerobes found in human infections and also lists some of the synonyms found in the literature. Appendix B illustrates colonial and Gram-stain morphology.

This monograph briefly surveys the role of anaerobic bacteria in infections and certain other pathophysiologic processes. Because the monograph is written primarily for clinicians and medical students, it highlights the clinical features of diseases caused by anaerobes, the clinical and bacteriologic diagnosis of anaerobic infections, and therapy. More in-depth information on clinical aspects of anaerobic infections and on anaerobic bacteriology can be found in the resources listed in the bibliography.

Clinical Features
of Diseases
Involving Anaerobes

ANAEROBIC BACTERIA IN PHYSIOLOGIC AND PATHOPHYSIOLOGIC PROCESSES

Anaerobic bacteria are an essential part of the body's normal flora. Surprisingly little is known about the nutritional and physiologic roles of the normal flora, particularly that of the intestinal tract; however, whatever the role, it is likely that anaerobic bacteria play an important part in view of their numerical dominance. For example, both an anaerobe, *Bacteroides fragilis*,* and a facultative organism, *Escherichia coli*, are involved in the intestinal synthesis of vitamin K; however, because *B fragilis* outnumbers *E coli* 1000 to 1 in the colon, *B fragilis* is likely to play a much more important role.

Various transformations of bile acids also result from bacterial activity in the intestine. Bile acids play essential roles in bile formation, fat absorption, and cholesterol metabolism. Anaerobic bacteria are known to be very active in deconjugating, dehydroxylating, and otherwise modifying bile acids. Although certain of these transformations are important for the conservation of bile acids by enterohepatic circulation, their full significance remains to be determined.

Although we are not able to document the protective effects of normal anaerobic flora in humans, data from mouse studies show that certain *Bacteroides* sp protect against infection with *Salmonella* and *Shigella* species. It is important to note, however, that colitis due to *Clostridium difficile* is seen almost entirely in individuals whose bowel flora has been altered by antimicrobial agents or by disease.

Even though anaerobic bacteria play beneficial roles, under certain circumstances they can also be pathogenic, invade tissue, and produce devastating disease. Table 1 lists types of infections in which anaerobes are commonly found or are the predominant pathogens.

One of the body's major defenses against infection by anaerobes is the normal oxidation-reduction potential (eH) of +120 mV. (Oxidation-reduction potential is also

---

Table 1.
Types of infections in which anaerobes are commonly found or are the predominant pathogens

Brain abscess
Otogenic meningitis, extradural or subdural empyema
Chronic otitis media
Dental and oral infections
Bite infections
Infections following head and neck surgery
Pneumonia secondary to obstructive process
Aspiration pneumonia
Lung abscess
Bronchiectasis
Thoracic empyema
Breast abscess
Liver abscess
Pylephlebitis
Peritonitis
Appendicitis
Subphrenic abscess
Other intra-abdominal abscesses
Wound infections following bowel surgery or trauma
Puerperal sepsis
Postabortal sepsis
Endometritis
Salpingitis
Tubo-ovarian abscess
Pelvic abscess
Other gynecologic infections
Perirectal abscess
Gas-forming cellulitis
Gas gangrene
Necrotizing fasciitis
Infected diabetic foot ulcers
Other soft-tissue infections
Osteomyelitis

---

*Whenever we speak of *Bacteroides fragilis*, we are referring to the *B fragilis* group, which includes a number of species previously designated as subspecies of *B fragilis*. Included among these are *B fragilis* itself, *B thetaiotaomicron*, *B vulgatus*, *B distasonis*, *B ovatus*, and *B uniformis*. *B fragilis* and *B thetaiotaomicron* are the species most commonly encountered clinically.

referred to frequently as redox potential.) A lower eH permits multiplication of anaerobes within tissues – even tissues such as teeth and lungs that are exposed to air. Lowered eH results from impaired blood supply, tissue necrosis, and growth of aerobes or facultative bacteria. Thus, vascular disease, epinephrine injection, cold, shock, edema, trauma, surgery, foreign bodies, malignancy, gas production by microorganisms, and aerobic infection significantly predispose tissue to anaerobic infection.

Certain hereditary, congenital, or acquired defects also predispose tissue to anaerobic infection. An example is acatalasia, a rare genetic defect characterized by the absence of the enzyme catalase, which results in recurrent infections of the gingivae and associated oral structures.

The toxins produced by some anaerobic bacteria are responsible for the virulence of certain anaerobic infections or intoxications. Examples are the toxins of *Clostridium tetani*, which cause tetanus, and of *C botulinum*, which causes botulism. Some toxins may be exzymes; for example, the alpha-toxin produced by *C perfringens* is a potent lecithinase that is hemolytic and necrotizing. The septic thrombophlebitis commonly seen in anaerobic infections may result from the production of heparinase by anaerobic bacteria – alone or in combination with other factors that accelerate coagulation; lysis of the clot can lead to metastatic abscesses. Other enzymes associated with anaerobic infection are collagenase, hyaluronidase, deoxyribonuclease, and the proteases.

BRAIN ABSCESS
Anaerobic bacteria such as *Peptostreptococcus* sp, *Fusobacterium* sp, and *Bacteroides* sp are responsible for most brain abscesses. Often, a mixture of anaerobes or a mixture of anaerobes and facultative organisms, or even fungi and other organisms, are responsible. Table 2 lists

anaerobes and other organisms involved in brain abscess.

The presence of anaerobes as normal flora in the oropharynx affords ample opportunity for infection in areas adjacent to it. Brain abscess can also develop after spread of infection from contiguous foci in the middle ear, mastoid, sinuses, and oropharynx, or by metastatic spread from the lungs. Right-to-left shunts in the heart or lungs also predispose to brain abscess. Table 3 lists conditions that predispose to brain abscess.

The clinical signs and symptoms of brain abscess (Table 4) are not as indicative of infection as they are of an intracerebral mass. Cerebrospinal fluid findings are variable: the protein level is usually elevated, the sugar is normal, and leukocytosis is absent or minimal. The brain scan (Figure 1) and computerized axial tomography (CT) scan are both useful diagnostic procedures, as are pneumoencephalography, EEG, arteriography, and echoencephalography.

Cerebral arteriography will frequently localize masses by demonstration of displacement or straightening of vessels, as shown in Figure 2, which illustrates a posteriorly located abscess. Needle aspiration of this abscess and instillation of a radiopaque dye provided an outline of its exact location (Figure 3). This procedure has rarely been necessary since the advent of the CT scan. An aerobic streptococcus and an anaerobic nonsporulating gram-positive bacillus were cultured from the aspirated material. The abscess was believed to have been caused by metastasis from a pulmonary arteriovenous fistula in the left lower lobe of the lung, as shown in Figure 4.

Treatment of brain abscess requires surgical drainage or excision of the space-occupying mass and appropriate antimicrobial therapy. In the absence of specific indentification of the organisms involved, treatment should include an antibacterial drug active against *Bacteroides fragilis*. Metronidazole or chloramphenicol would be the drug of choice in this situation. Clin-

Table 2.
Organisms involved in brain abscess

Anaerobes
Bacteroides fragilis
Other *Bacteroides* sp
Fusobacterium (especially *F necrophorum)*
Anaerobic cocci and streptococci
Actinomyces israelii
Other gram-positive nonsporulating rods
Clostridium

Facultative or microaerophilic organisms
Microaerophilic cocci and streptococci
Pneumococcus
Viridans streptococci
Staphylococcus aureus
Haemophilus influenzae
Haemophilus aphrophilus
Other gram-negative bacilli
Nocardia asteroides
Meningococcus

Miscellaneous organisms
Mycobacterium sp
Fungi
Entamoeba histolytica

Table 3.
Conditions predisposing to brain abscess

Head injury
Intracranial surgery
Chronic otitis media, mastoiditis
Sinusitis – particularly frontal and sphenoidal sinuses
Infections of face and scalp
Tonsillitis
Lung abscess, pneumonia, empyema, bronchiectasis
Congenital heart disease with right to left shunt
Bacterial endocarditis
Sepsis, particularly that following abortion or dental extraction
Carbuncles

Table 4.
Clinical signs and symptoms of brain abscess

Low-grade to moderate fever – or no fever
Headache
Drowsiness
Confusion, stupor
Seizures – generalized or focal
Nausea, vomiting
Focal motor, sensory, or speech disorders
Papilledema
Leukocytosis
Evidence of primary focus of infection

Figure 1. Brain scan demonstrating a mass (in this case an abscess) in the occipital region, left of the midline. The mass caused an increased uptake of radioisotope.

Figure 2. Cerebral arteriogram demonstrating displacement and "straightening" of vessels by a posteriorly located brain abscess.

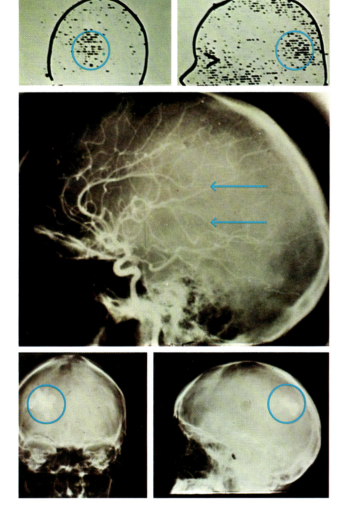

Figure 3. Radiograph of the same abscess, outlined by the injection of radiopaque dye following needle aspiration of the abscess.

damycin is also quite active in vitro against *B fragilis;* however, clindamycin does not readily cross the blood-brain barrier and is, therefore, not indicated for the treatment of brain abscess. Encapsulated abscesses (Figure 5) can sometimes be removed intact. In other instances, needle aspiration of the abscess plus antimicrobial therapy are successful forms of treatment.

Clinicians should be aware that other intracranial lesions, such as meningitis and subdural or epidural empyema, may involve anaerobic bacteria.

EAR, NOSE, THROAT, HEAD, AND NECK INFECTIONS

Because of the widespread use of penicillin and other antibiotics that are active against anaerobes, these organisms are less frequently recovered from otitis media, mastoiditis, tonsillitis, and sinusitis than they were before the availability of such drugs; often the infections are cured before they become chronic or complicated. Nevertheless, anaerobes may play an important role in otitis media, mastoiditis, and sinusitis, particularly if the infections are chronic.

So-called Vincent's angina, a membranous tonsillitis caused primarily or exclusively by *Fusobacterium necrophorum,* is occasionally complicated by sepsis and widespread metastatic infection (eg, to bone, joints, lung, liver). Peritonsillar abscesses and wound infections following radical head and neck surgery for cancer also commonly involve anaerobes.

The role of anaerobes in Ludwig's angina is discussed in the section on Oral Infections.

ORAL INFECTIONS

Anaerobic bacteria inhabit the oral region as normal flora and are abundant in the gingival sulci. They are also present in saliva and mature dental plaque (microbial aggregations attached to teeth). When facultative bacteria, particularly streptococci, adhere to the surfaces

Figure 4. Radiograph demonstrating a left lower lobe pulmonary arteriovenous fistula that was implicated in the development of the brain abscess shown in Figure 3.

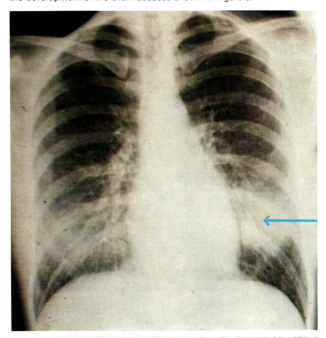

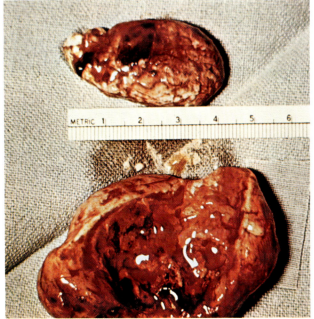

Figure 5. Two encapsulated brain abscesses. The neurosurgeons were able to remove these abscesses intact.

of the teeth and lower the oxidation-reduction potential (eH), anaerobic organisms can flourish, even at sites exposed to large amounts of air (Figures 6A and 6B).

The role of anaerobes in infections within the oral cavity is well established, and sites of intraoral infections are shown in Figure 7. Anaerobic bacteria normally found in the oral cavity can also cause infections elsewhere in the body, as shown in Figure 8. Oral anaerobes found in such infections are listed in Table 5. Even though oral anaerobes are responsible for a wide range of infections, this section will discuss only intraoral infections.

Root Caries

Although many anaerobes are found in mature dental plaque, their role in dental decay is not well defined. (Anaerobes do not cause tooth destruction in animal models.) While it is true that root caries occur in a low eH environment, oxygen-fastidious anaerobes are not numerous at the site. Recent studies suggest that *Actinomyces viscosus* or *A naeslundii* may be the causative agents in this type of infection.

Root Canal and Periapical Infections

Root canal infections commonly occur after ingress of bacteria, usually through carious lesions or by way of disease processes in the supporting tissues of the teeth. Facultative streptococci are most frequently recovered on culture, although in recent carefully conducted studies, anaerobes were recovered in 60% or more of the cultures. Infected root canals and periapical abscesses are frequently the source of bacteremia and a resultant bacterial endocarditis. Although facultative streptococci are the principal isolates, anaerobic sepsis is common, so anaerobic blood cultures should always be obtained in cases of suspected bacteremia.

Figures 6A and 6B. Before (6A) and after (6B) application of the redox indicator benzyl viologen; the stained areas indicate the presence of extremely anaerobic conditions (eH of −360 mV or below). Areas exposed to air can become anaerobic as a result of metabolic activity of facultative organisms.

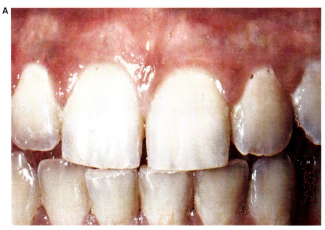

A

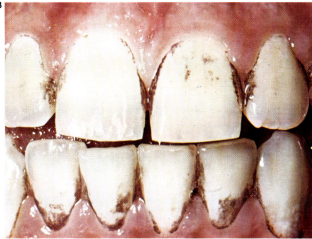

B

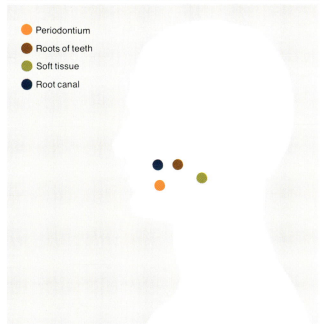

- 🟠 Periodontium
- 🟤 Roots of teeth
- 🟢 Soft tissue
- 🔵 Root canal

Figure 7. Sites of intraoral anaerobic infections.

Figure 8. Major locations of anaerobic infections caused by oral anaerobes.

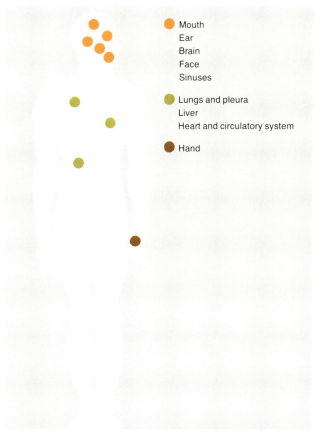

- 🟠 Mouth
  Ear
  Brain
  Face
  Sinuses

- 🟢 Lungs and pleura
  Liver
  Heart and circulatory system

- 🟤 Hand

Periodontal Disease

Surveys show that it is likely that every adult in North America has some degree of periodontal disease, and half of those who retain teeth to age 50 have gross and extensive periodontal tissue destruction. There are several types of periodontal diseases, the most common being gingivitis and chronic periodontitis. Gingivitis involves gingival irritation without bone loss. Chronic periodontitis, however, involves gingivitis with deep gingival pockets and loss of supporting bone. Streptococci and filamentous organisms, chiefly *Actinomyces viscosus* and *A naeslundii*, adhere to teeth to form plaque and thereby lower the eH. Many gram-negative anaerobes and spirochetes multiply in the crevices between teeth and gums (gingival sulci). The bacteria accumulate in large masses and initiate a destructive process. Possible mechanisms of destruction are the production of lytic enzymes such as chondroitinase, collagenase, hyaluronidase, and proteases; cytotoxins and endotoxins; or the initiation of inflammation through a hypersensitivity reaction.

14

Table 5.
Indigenous oral anaerobes found in nonoral infections

(Only one representative species is listed)

Actinomyces israelii
Arachnia propionica
Bacteroides melaninogenicus
Bifidobacterium eriksonii
Fusobacterium nucleatum
Peptococcus asaccharolyticus
Peptostreptococcus anaerobius
Propionibacterium acnes
Treponema vincentii
Veillonella parvula

Examples of acute and rapidly destructive periodontal disease are juvenile periodontitis (periodontosis) and acute necrotizing ulcerative gingivitis (ANUG). Periodontosis is seen in persons up to 25 years of age and is characterized by minimal gingivitis and plaque formation and by rapid destruction of the supporting bone. Five groups of gram-negative anaerobic bacteria have been implicated in this disease, although they presently cannot be classified. ANUG is seen most commonly in persons between adolescence and age 30. Early lesions are distinguished by necrosis and ulceration of the interdental gingiva, with bleeding, extreme pain, and development of a pseudomembrane. Large spirochetes and anaerobic fusiform bacilli are seen in exudates from diseased areas; some investigators believe that the large spirochetes play an important role because they are seen in ANUG but not in other types of gingivitis or periodontal disease.

Oral Soft-Tissue Infections
Oral anaerobic soft-tissue infections include stomatitis, Ludwig's angina, dental and oral abscesses, phlegmons, and actinomycosis.

Abscesses and Phlegmons: Dental and oral abscesses and phlegmons usually are caused by *Bacteroides* and anaerobic cocci, or involve both anaerobes and facultative organisms. Putrid odor, inflammation, and frank pus are the primary clinical signs. These infections can arise after dental manipulation or after trauma inflicted by toothpicks or toothbrushes, but they can also occur with no history of injury.

Stomatitis: Stomatitis is usually an extension of untreated ANUG; it destroys mucous membranes and may even penetrate the adjacent facial tissue and produce **noma.** Many kinds of organisms have been found in stomatitis, the most common being *Fusobacterium* sp, spirochetes, *B melaninogenicus,* * anaerobic cocci, and clostridia.

*Originally, *Bacteroides melaninogenicus* was made up of three subspecies. One of these, *B asaccharolyticus*, is now accorded full species status. Accordingly, when we speak of *B melaninogenicus*, we are referring to a "*B melaninogenicus-B asaccharolyticus*" group. Additional species of black pigmented bacteroides, such as *B gingivalis*, are being recognized. In the past, these would have been classified as *B melaninogenicus*.

Ludwig's Angina: Ludwig's angina (Figure 9) is an infectious process involving the submaxillary and sublingual spaces. The disease runs a rapid and violent course. Edema is marked, but there is little or no pus. The edema causes great difficulty in swallowing, and a tracheotomy or intubation is frequently necessary to permit respiration. Oral anaerobic bacteria are frequently involved in this infection, which can occur after dental manipulation.

Actinomycosis: Actinomycosis of the mouth or face usually involves the obligate anaerobe, *Actinomyces israelii.* Less commonly, other species of *Actinomyces,* such as *A naeslundii,* or *Arachnia propionica,* which are present as normal flora in the mouth, may be found in actinomycosis, an infection that may involve periodontal tissues, the oropharynx, neck, jaw, and the lower respiratory tract as well as, less commonly, the abdomen and the central nervous system. These infections are characterized by chronic firm induration, swelling, and sinus and fistula formation (Figures 10-12). Pulmonary actinomycosis may extend to the pleural space, the ribs, and the chest wall. Exudate from the sinuses may contain typical "sulfur granules" (Figures 13-14), which after Gram staining can be seen to consist of filamentous organisms (Figure 15). Actinomycosis is typically a mixed infection involving, in addition to *Actinomyces* or *Arachnia, B melaninogenicus, Actinobacillus actinomycetemcomitans,* and other organisms. Actinomycosis can disseminate hematogenously from a primary lesion to produce abscesses throughout the body. Penicillin is the drug of choice for treating these infections, and prolonged treatment (often for months) may be necessary to prevent relapse.

LUNG AND PLEURAL SPACE INFECTIONS

Pleuropulmonary infections caused by anaerobes are relatively common but are overlooked more frequently

Figure 9. Ludwig's angina. Massive edema, now subsided, necessitated tracheotomy.

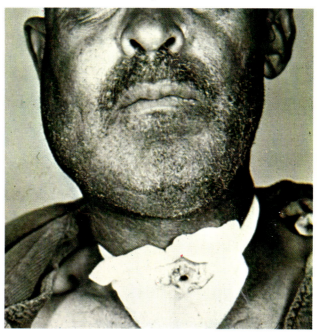

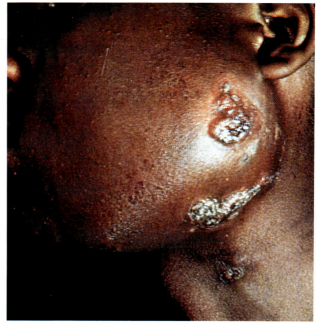

Figure 10. Oral actinomycosis frequently leads to draining sinuses of jaw.

Figure 11. Actinomycosis of the neck. Facial lesions frequently spread to involve the neck. Multiple fistulae are common with this disease.

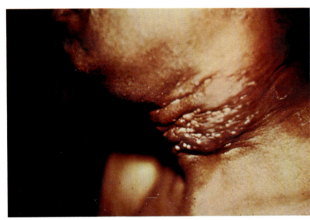

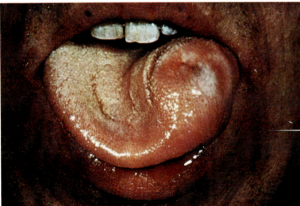

Figure 12. Actinomycotic lesion of the tongue. *Actinomyces israelii* is an important organism in actinomycosis, an infection that can occur if this indigenous organism enters the tissue following trauma.

Figure 13. Gross appearance of "sulfur granules." These are actually colonies of *Actinomyces* or *Arachnia*, which grossly resemble granules of sulfur.

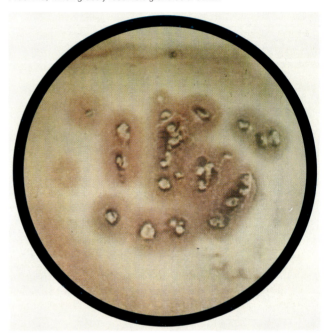

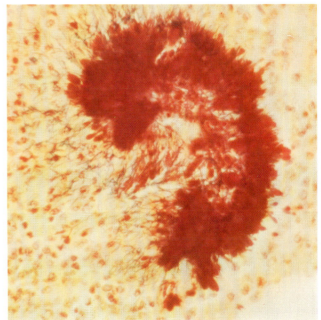

Figure 14. Sulfur granule (bacterial aggregation) in tissue.

Figure 15. Histologic section of mandible with
actinomycotic infection; thin filamentous organisms
are present in necrotic tissue adjacent to the
intact bone.

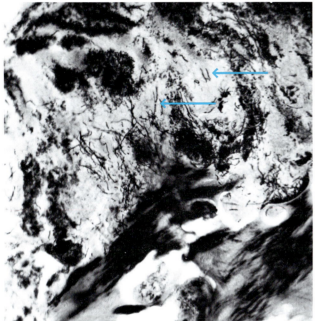

than are other types of anaerobic infections. There are
several reasons for this: **1** clinicians lack knowledge of
the frequency, importance, and clinical characteristics
of these infections; **2** sputum specimens are unsuitable
for anaerobic culture; **3** there is no characteristic foul
odor of sputum during the first few days of parenchymal
infection (early administration of antibiotics may also
be responsible for the absence of this symptom, and 50%
of patients never develop foul-smelling sputum or empyema fluid); and **4** bacteremia is uncommon in this
type of infection.

Conditions commonly predisposing to anaerobic
pleuropulmonary infection are listed in Table 6, and
clues to the presence of these infections are listed in
Table 7. Sputum is not suitable for anaerobic culture
because it contains large numbers of many species of
anaerobes from the indigenous flora of the mouth and
pharynx. Specimens of material properly obtained by
percutaneous transtracheal aspiration (Figure 16), how-

Table 6.
Conditions predisposing to anaerobic pleuropulmonary infection

Aspiration
Altered consciousness
   Alcoholism
   General anesthesia
   Drug overdose
   Other cause for aspiration
Esophageal dysfunction or intestinal obstruction
Tonsillectomy or tooth extraction

Preceding extrapulmonary anaerobic infection
Periodontal disease
Pharyngeal infection
Otitis, mastoiditis
Female genital tract infection
Intra-abdominal infection
Bacterial endocarditis

Postthoracotomy or penetrating chest wound

Local conditions
Bland (sterile) pulmonary infarction
Bronchogenic carcinoma
Bronchiectasis
Foreign body

Systemic conditions
Diabetes mellitus
Nonpulmonary malignancy
Corticosteroid or immunosuppressive therapy

Table 7.
Clues to presence of anaerobic pleuropulmonary infection

Alcoholic patient

Subacute or chronic onset and progression of disease (but may be acute or fulminant)

Fetid discharge

Tissue necrosis (such as abscess formation or bronchopleural fistula)

Gram stain of clinical specimen exhibiting morphology suggestive of anaerobes

Failure to recover a likely pathogen on aerobic culture

Presence of a predisposing condition (see Table 6)

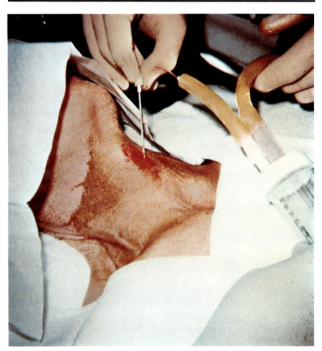

Figure 16. Method of collection of specimen by transtracheal aspiration. This technique avoids contamination with oral anaerobes.

ever, are reliable for anaerobic culture and are an important source for diagnosis of anaerobic lung infections. Other specimens suitable for diagnostic purposes are blood, pleural fluid (in patients with empyema), material obtained by percutaneous transthoracic aspiration or by thoracotomy, and material obtained from other body sites that may be involved in the same infectious process. It is not yet established that material obtained via a bronchoscope, using a bronchial brush within a special double-lumen catheter with a distal polyethylene glycol plug, is a reliable specimen source. Transport of these specimens – which are often small – under anaerobic conditions is absolutely crucial. (Details on the transport of anaerobic specimens are discussed under Diagnosis and Bacteriology of Anaerobic Infections.)

Complete identification of the anaerobes present is desirable to ensure proper therapy because many antibiotics are not effective against *Bacteroides fragilis*, and this organism is found in 5% to 10% of anaerobic pleuropulmonary infections. Other beta-lactamase-producing anaerobes, chiefly gram-negative rods, may also be found in 5% to 10% of such infections, and other drug-resistant anaerobes may be encountered. Surprisingly, in pleuropulmonary infection, *B fragilis* originates from infection sites above the diaphragm as frequently as it does from intraperitoneal sites. Presumably, *B fragilis* may be part of the flora of the oropharynx, the nasopharynx, or gastric contents of certain individuals under certain circumstances.

Most anaerobic pleuropulmonary infections involve two to nine species of anaerobes as well as aerobes or facultative organisms. In cases in which a single anaerobe is recovered in pure culture, the most common isolates are *F necrophorum*, *F nucleatum*, *Peptostreptococcus* sp, and *Clostridium perfringens*. *F necrophorum* infections, once common following tonsillitis (Vincent's angina) or tonsillectomy, are seen much less frequently since the advent of antimicrobial agents.

Treatment of anaerobic infections of the lung and pleura involves surgical drainage of empyema, debridement of necrotic tissue (when this is present), and antimicrobial therapy. Lung abscesses almost never require surgery. Resolution of anaerobic pleuropulmonary infections is typically very slow, averaging nine weeks for a solitary abscess, 20 weeks for necrotizing pneumonia, and 34 weeks for empyema. Prolonged and adequate antimicrobial therapy is indicated to minimize the possibility of a relapse or any residual damage.

Patients with anaerobic pleuropulmonary infection require close follow-up because these infections are sometimes the first clue to the presence of a malignancy, and occasionally they metastasize to the brain.

Even with proper antimicrobial therapy, the mortality in anaerobic pleuropulmonary infections is 15%. Mortality is greatest among patients with necrotizing pneumonia or with serious underlying conditions and in those who receive inappropriate therapy.

Specific anaerobic infections of the lung and pleural space include pneumonia (acute or chronic), aspiration pneumonia (particularly in alcoholic patients), necrotizing pneumonia, septic pulmonary infarction, lung abscess, empyema, and pulmonary actinomycosis. These infections are often misdiagnosed as tumors when the disease is subacute or chronic, because of marked weight loss of the patient, the indolent course of the disease, and the occasional presence of mediastinal adenopathy. These infections will be reviewed here.

Pneumonia

Radiographic features of uncomplicated pneumonia, a common anaerobic pleuropulmonary infection, are shown in Figure 17. Infiltrates are present in the left upper lobe and in the posterior basal segment of the left lower lobe. This 44-year-old man had left chest pain, fever, and cough productive of yellow-green sputum. Symptoms had been present for seven days. Ten months

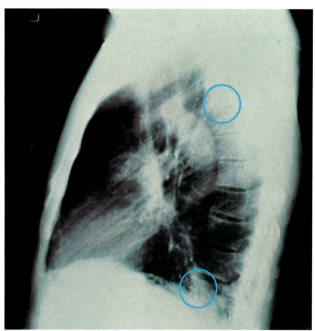

Figure 17. Pneumonia. Lateral radiograph shows infiltrates in the left upper lobe and in the posterior segment of the left lower lobe.

Figures 18A-18G. Pneumonia.

Figures 18A and 18B. PA and lateral radiographs show pneumonia in superior segment of left lower lobe and a prominent left hilum. Pneumonia stemmed from dental infection or sinusitis and metastasized to the brain.

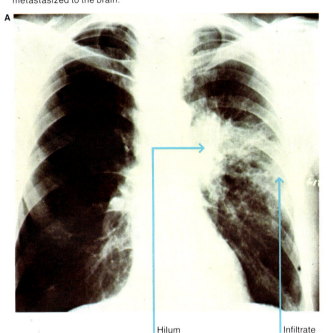

Hilum   Infiltrate

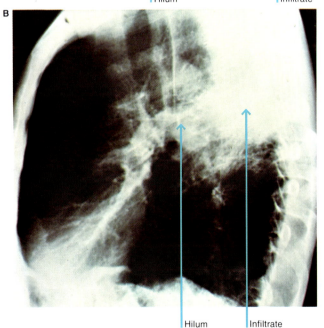

Hilum   Infiltrate

previously, this patient had been admitted with similar symptoms and identical infiltrates. He was not treated at that time, however, because his symptoms had improved spontaneously. With the present episode, transtracheal aspirate yielded *B fragilis, B melaninogenicus, Fusobacterium* sp, *P acnes, Peptococcus* sp, and *H influenzae*, all in counts $> 10^6$/ml. This indolent pneumonia without cavitation cleared with one month of chloramphenicol therapy.

Figures 18A-18G represent a rather involved case of pneumonia that may have originated from periodontal disease or sinusitis and then metastasized to the brain. Figures 18A and 18B show an area of pneumonia in the superior segment of the left lower lobe and a prominent left hilum. This 52-year-old diabetic male had emphysema, a history of sinusitis with foul-smelling postnasal discharge, and poor dental health. Figures 18C and 18D are radiographs of his teeth. The first shows evidence of some periodontal disease and root caries. The second, taken 15 months later, shows evidence of periapical abscesses in the right lower jaw and progressive periodontal disease and root caries.

Suspecting an anaerobic infection, the ward physician obtained a transtracheal aspiration specimen for culture, which was negative for anaerobes. Unfortunately, a Gram stain was not done. The physician began a work-up for lung cancer. Meanwhile, focal seizures developed and a cerebral lesion (abscess) was found; during brain surgery, 40 ml of foul-smelling pus was drained. A Gram-stained smear of the pus is shown in Figure 18E. Delicate pale-staining, filamentous, gramnegative bacilli proved to be *F nucleatum* on culture. Gram stain of a second transtracheal aspiration (Figure 18F) showed organisms very much like those obtained from the brain abscess. A culture, however, was negative for anaerobes — probably because the patient had been started on penicillin therapy.

Figure 18C. Dental radiograph shows evidence of periodontal disease and root caries in the lower jaw.

Figure 18D. Dental radiograph taken 15 months later shows evidence of periapical abscesses in the right lower jaw and progressive periodontal disease and root caries.

Figure 18F. Gram stain of transtracheal aspirate shows organisms very much like those found in the brain abscess.

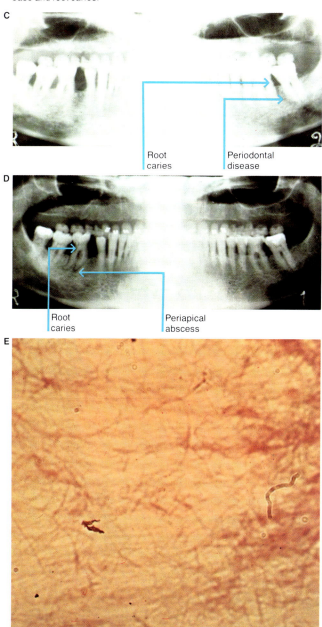

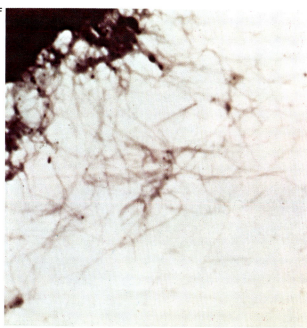

Figure 18E. Gram stain of specimen from brain abscess. The pale-staining delicate filamentous gramnegative bacilli proved to be *F nucleatum* on culture.

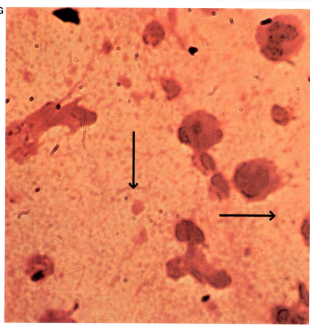

Figure 18G. Gram stain of transtracheal aspirate obtained from the same patient during a later episode of pneumonia. Pale gram-negative bacilli again proved to be *F nucleatum* on culture.

Figures 19A and 19B. Aspiration pneumonia. PA and lateral radiographs show circumscribed oval density in posterior segment of right upper lobe.

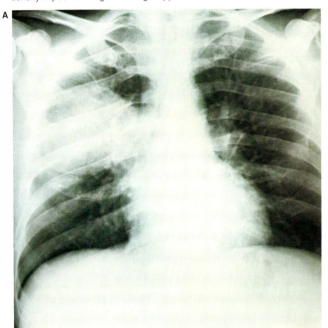

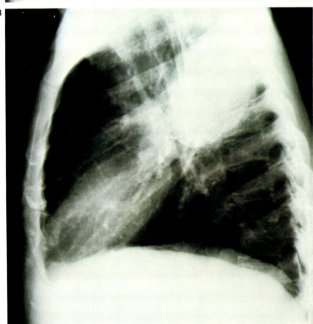

This patient was treated for many months (antibiotic treatment on an outpatient basis following intensive penicillin therapy in the hospital). Dental treatment was initiated, and his chest returned to normal. Almost a year later, this patient returned with another episode of pneumonia, this time in the posterior segment of the left upper lobe. A Gram stain of transtracheal aspirate (Figure 18G) showed a few pale gram-negative bacilli. *F nucleatum* again grew in pure culture, and the infection again responded well to antibiotic therapy.

Aspiration Pneumonia

Radiographic features of aspiration pneumonia, commonly associated with alcoholism, are shown in Figures 19A and 19B. A circumscribed oval density is present in the posterior segment of the right upper lobe in this patient. Aside from the location, which is a common site for aspiration since it is a dependent segment, there is nothing to suggest the presence of anaerobic infection. Diagnosis of anaerobic pneumonia was established after organisms obtained by transtracheal aspiration were identified.

Figures 20A-20E summarize a case of fairly extensive aspiration pneumonia. This 44-year-old alcoholic was hospitalized with complaints of cough productive of foul-smelling sputum, pleurisy, and weakness. Symptoms had been present for one week. WBC was 25,700/cu mm. A sputum specimen yielded two potential pathogens — *S pneumoniae* and *H influenzae* — on culture, whereas a specimen of transtracheal aspirate yielded a pure culture of *Peptostreptococcus* sp. Figures 20A and 20B show chest roentgenograms, with pneumonia in the superior segment of the right lower lobe. Figure 20C, a radiograph taken five days after the patient was hospitalized, shows a large air-fluid level and a multiloculated cavity, and figure 20D shows clearing of the infection in response to therapy. Figure 20E is a graphic representation of the course of infection and therapy.

Figures 20A-20E. Aspiration pneumonia.

Figures 20A and 20B. PA and lateral radiographs show fairly extensive pneumonia in superior segment of the right lower lobe.

Figure 20C. PA radiograph taken 5 days later shows large air-fluid level. Cavity is multiloculated.

A
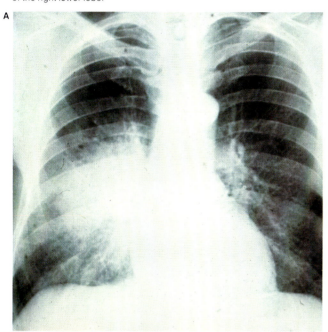

C
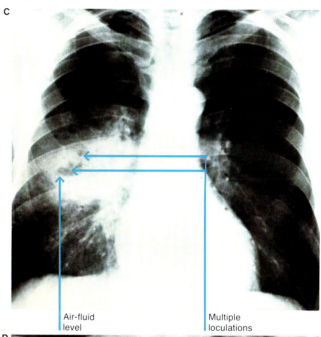

Air-fluid level

Multiple loculations

B

D
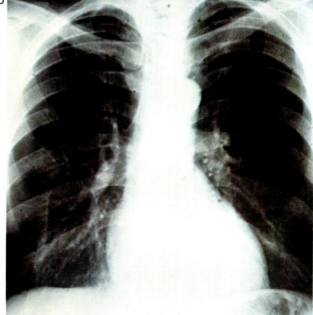

Figure 20D. Follow-up radiograph shows clearing of infection.

Figure 20E. Course of infection and therapy.

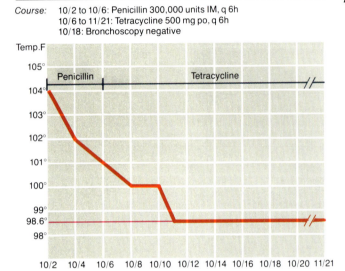

Course: 10/2 to 10/6: Penicillin 300,000 units IM, q 6h
10/6 to 11/21: Tetracycline 500 mg po, q 6h
10/18: Bronchoscopy negative

Patient discharged asymptomatic 10/21

Figures 21A-21C. Aspiration pneumonia with empyema.

Figures 21A and 21B. PA and lateral radiographs show multiloculated empyema resulting from aspiration of vomitus.

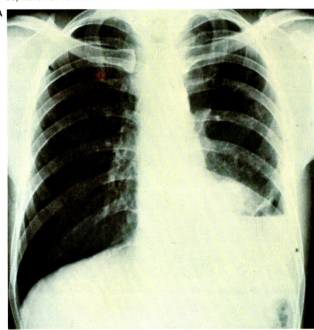

A

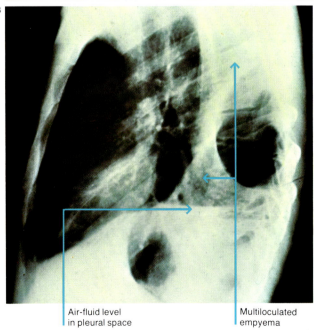

B

Air-fluid level
in pleural space

Multiloculated
empyema

Figure 21C. Gram stain of empyema specimen.
Culture yielded *B melaninogenicus, B fragilis,
B oralis, Fusobacterium* sp, *Peptococcus* sp,
Peptostreptococcus sp, *Eubacterium* sp, and
Propionibacterium acnes.

Figures 22A and 22B. Necrotizing pneumonia.
PA and lateral radiographs show significant infiltrate
and air-fluid level in superior segment of the left
lower lobe and infiltrate at the base in a posterior
location.

C

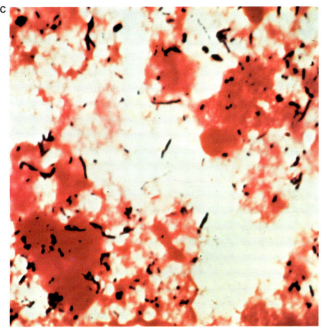

A

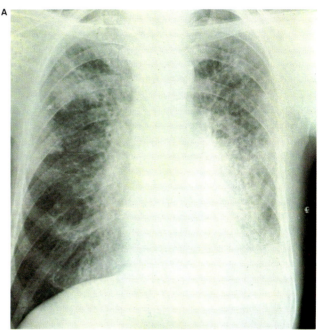

B

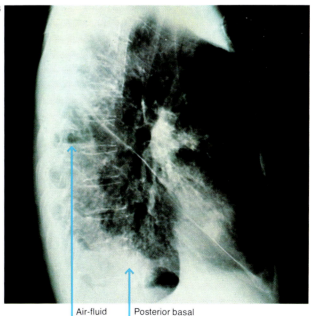

Air-fluid
level

Posterior basal
infiltrate

Figures 21A and 21B show radiographic features of
aspiration pneumonia with multiloculated empyema
that resulted from aspiration of vomitus by a 25-year-
old man who had attempted suicide with barbiturates.
The patient was intubated and, because of the growth
of *Klebsiella* sp on culture of a sputum specimen, was
treated with gentamicin. This patient was later trans-
ferred to Wadsworth VA Hospital. Radiographs of his
chest showed the presence of multiloculated empyema,
which was drained surgically (open drainage). The
patient was treated with chloramphenicol, then with
clindamycin, and the infection cleared. Figure 21C
shows a Gram stain of a specimen of this foul-smelling
empyema. Culture yielded *B melaninogenicus, B fragilis, B oralis, Fusobacterium* sp, *Peptococcus* sp, *Peptostreptococcus* sp, *Eubacterium* sp, and *Propionibacterium acnes.*

Figure 23. Necrotizing pneumonia with empyema. PA radiograph shows cavitary infiltrate in the right upper lobe and small empyema at the right base.

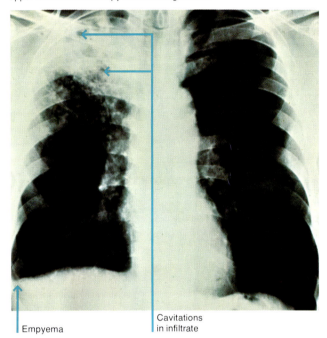

Empyema

Cavitations in infiltrate

Necrotizing Pneumonia

Necrotizing pneumonia is a suppurative inflammatory process with tissue destruction leading to multiple small cavities. Figures 22A and 22B show radiographic features of an extensive infection of this type with multiple abscesses in the superior segment of the left lower lobe. An infiltrate is also present at the left base. Figure 23 is the chest radiograph of another patient who had necrotizing pneumonia and empyema. A cavitary infiltrate is present in the right upper lobe, and a small empyema is present at the right base. This 57-year-old alcoholic patient was hospitalized with complaints of malaise (present for five months) and a cough productive of purulent sputum (present for nine days). Findings from tuberculosis and tumor work-ups were normal. Specimens of sputum and bronchial washings yielded *Neisseria* and alpha-hemolytic streptococci on culture; however, culture of a transtracheal aspirate specimen yielded *Fusobacterium* sp and *Peptostreptococcus* sp. This infection responded to medication, and subsequent chest radiographs showed that the lungs had become normal. Three years later, a bronchogenic carcinoma developed in the same area of lung and the patient died. Presumably it was present earlier but was too small to be detected.

Radiographs in Figures 24A and 24B show two stages of necrotic pneumonitis that developed in the left upper and lower lobes. Empyema was also present at the right base. This 61-year-old alcoholic patient complained of a cough productive of foul-smelling sputum. A sputum specimen yielded *Pseudomonas aeruginosa* on culture, but transtracheal aspirate yielded *Fusobacterium* sp, *B melaninogenicus*, and aerobic alpha-hemolytic streptococcus. This infection responded to therapy initially, but the patient suddenly became worse and died.

Septic Pulmonary Infarction

Septic pulmonary infarction can lead to cavitation and

Figures 24A and 24B. Necrotizing pneumonia with empyema.

Figure 24A. PA radiograph shows necrotizing pneumonia in left upper and lower lobes and empyema at right base.

Figure 25. Septic pulmonary infarction. Radiograph shows multiple nodules, some with cavitation, and left pleural effusion.

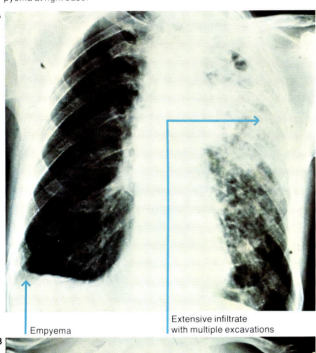

Empyema

Extensive infiltrate with multiple excavations

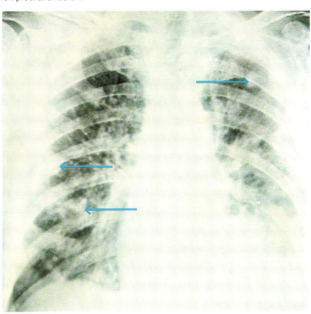

pleural effusion; however, multiple nodules are most characteristic of this disease process. Although the infection can result from foci such as subcutaneous or peritonsillar abscesses (by means of bacteremic seeding), it usually results from right-sided bacterial endocarditis or septic thrombophlebitis. Infected thrombi become lodged in the pulmonary circulation. Figure 25 shows radiographic features of septic pulmonary infarction with multiple nodules, cavitation, and left pleural effusion.

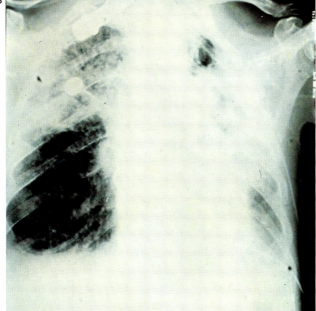

Figure 24B. Worsening of necrotizing pneumonia. Patient died.

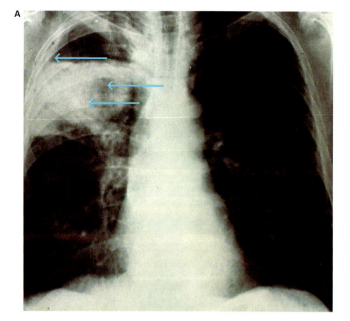

Figures 26A-26C. Lung abscess.

Figures 26A and 26B. PA and lateral radiographs show large multiloculated abscess cavity in posterior segment of right upper lobe and multiple air-fluid levels.

Figure 26C. Gross appearance of necrotic lung at autopsy.

Lung Abscess

Anaerobic pleuropulmonary infections frequently lead to abscess formation, and solitary lung abscess is primarily an anaerobic infection. Figures 26A and 26B show a large multiloculated abscess cavity in the posterior segment of the right upper lobe; there are multiple air-fluid levels. This 55-year-old alcoholic patient complained of cough, putrid sputum, fever, and weight loss. Symptoms had been present for three months. The patient died shortly after being hospitalized (of causes unrelated to the pulmonary infection), and at autopsy nine different species of anaerobes were recovered from material aspirated from the abscess. Figure 26C shows the appearance of the necrotic lung at autopsy.

Figure 27 shows an abscess with air-fluid level in the left upper lobe of a patient who was found to have a bronchogenic carcinoma. The necrotic center of this tumor had developed into an abscess. Similar infections may occur distal to segments obstructed by such tumors.

Empyema

Figure 28A shows leakage of gastric contents through a gastropleural fistula that developed in a 65-year-old man after an esophagogastrectomy for carcinoma of the stomach had been performed; Figure 28B shows extensive loculated empyema that developed as a result of the leakage; and Figure 28C summarizes the course of the polymicrobic pleural empyema and prolonged *Bacteroides* bacteremia that resulted from the leakage.

Figure 29 is an illustration of a massive putrid empyema in the right thorax. A culture of the aspirate yielded *Fusobacterium* sp, *Peptostreptococcus* sp, and microaerophilic streptococci. Open drainage and penicillin G therapy effected a cure.

Pulmonary Actinomycosis

Pulmonary actinomycosis tends to involve primarily the lower lobes. Consolidation occurs early and com-

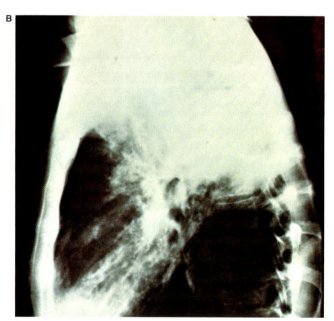

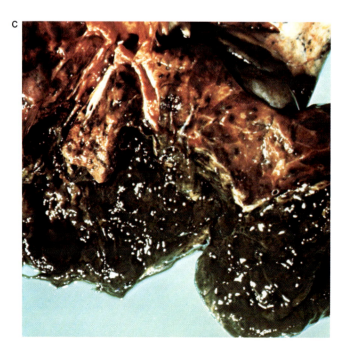

Figure 27. Abscess. Radiograph shows abscess with air-fluid level in left upper lobe. Patient was found to have bronchogenic carcinoma. The abscess was in the necrotic center of the tumor.

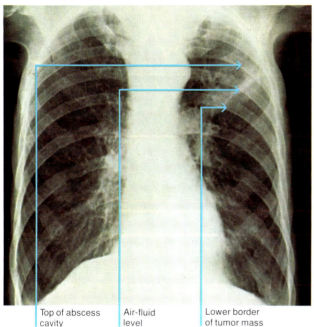

Top of abscess cavity | Air-fluid level | Lower border of tumor mass

Figures 28A-28C. Empyema.

Figure 28A. Radiograph of upper gastrointestinal area shows leakage of gastric contents through a gastropleural fistula.

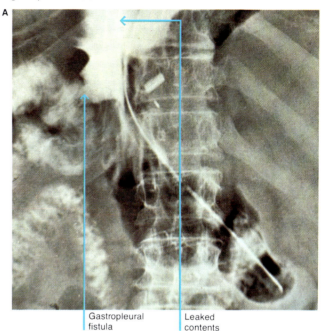

Gastropleural fistula | Leaked contents

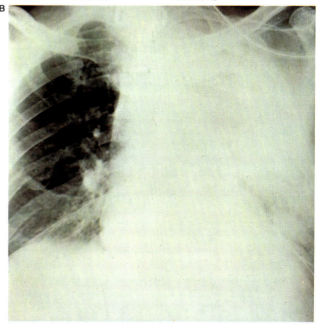

Figure 28B. PA radiograph shows extensive loculated empyema that developed as a result of the leakage.

monly extends to the pleural space and chest wall. Periostitis or osteomyelitis of the ribs and draining sinuses may also develop. Figure 30 shows a case of pulmonary actinomycosis that involved the entire left lung. Multiple abscesses are present and there is considerable loss of lung volume. As with other anaerobic pulmonary infections, actinomycosis tends to metastasize to the brain or vertebral column. Sulfur granules may be noted in discharges and are particularly evident if the discharged matter is filtered through gauze. These granules, which are colonies of *Actinomyces* or *Arachnia*, may be washed with sterile fluid and then cultured anaerobically.

Figure 28C. Summary of the course of infection.

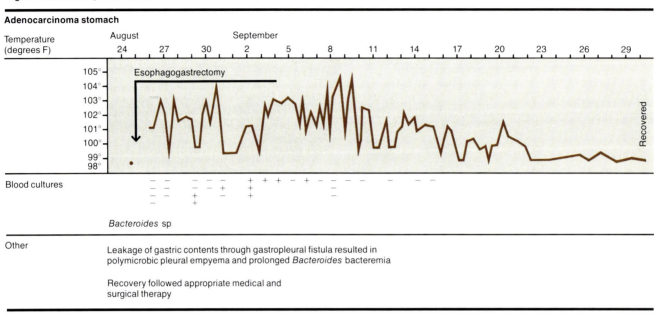

Adenocarcinoma stomach

Temperature (degrees F)

Blood cultures

Bacteroides sp

Other

Leakage of gastric contents through gastropleural fistula resulted in polymicrobic pleural empyema and prolonged *Bacteroides* bacteremia

Recovery followed appropriate medical and surgical therapy

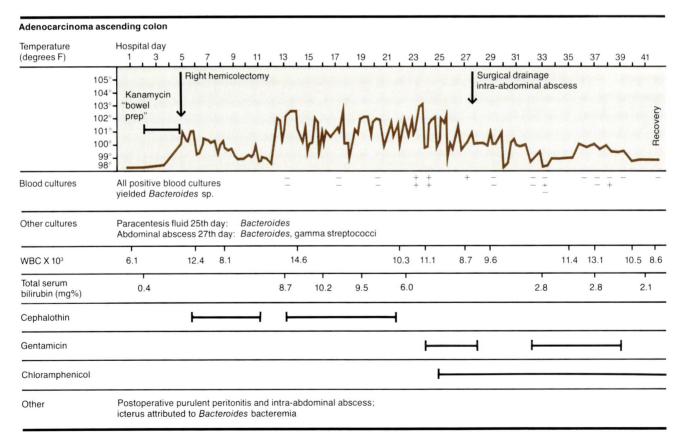

Adenocarcinoma ascending colon

Temperature (degrees F)

Blood cultures — All positive blood cultures yielded *Bacteroides* sp.

Other cultures — Paracentesis fluid 25th day: *Bacteroides*
Abdominal abscess 27th day: *Bacteroides*, gamma streptococci

WBC X 10³

Total serum bilirubin (mg%)

Cephalothin

Gentamicin

Chloramphenicol

Other — Postoperative purulent peritonitis and intra-abdominal abscess; icterus attributed to *Bacteroides* bacteremia

Figure 31. Summary of the course of *Bacteroides* bacteremia that developed in a patient who had undergone a hemicolectomy for a bowel tumor.

BACTEREMIA

An accurate estimate of the incidence of bacteremia caused by anaerobes is probably not available; however, a 2% to 10% incidence is presently being reported. Continued improvement in anaerobic blood culture methods and techniques of identification will undoubtedly show that the actual incidence is greater. The non-sporeforming gram-negative rods (collectively referred to as *Bacteroides* in most series) are the anaerobes most frequently recovered from blood cultures, whereas the clostridia and anaerobic cocci are found less often. At present, gastrointestinal and gynecologic infections are the primary sources of these bacteremias. Less common sources are pulmonary, oropharyngeal, paranasal, and sinus infections; infected decubitus ulcers; and wounds. In addition, patients with underlying disorders such as diabetes or malignancy, or those who have been treated with corticosteroids, will be more likely to develop anaerobic bacteremia.

The clinical picture of bacteremia consists of sudden onset of fever, rigors, diaphoresis, and often jaundice. Septic thrombophlebitis of pelvic veins or other veins, with subsequent embolization, is a serious complication. *Fusobacterium* possesses a potent endotoxin, but *Bacteroides* does not have a classic endotoxin (although weak endotoxic activity is present); nevertheless, septic shock and disseminated intravascular coagulation have occurred with *Bacteroides* bacteremia.

Figure 31 graphically summarizes a course of *Bacteroides* bacteremia that developed in a patient with adenocarcinoma of the bowel who had undergone a hemicolectomy. The patient first received cephalothin, then gentamicin, but did not improve until an intra-abdominal abscess was drained and chloramphenicol therapy was instituted. It is of interest that this patient received oral kanamycin preoperatively, because the aminoglycosides are generally inactive against anaerobes; kanamycin therapy likely decreased only the

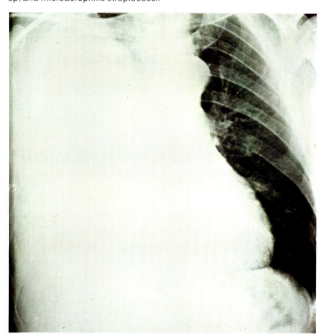

Figure 29. Empyema. Radiograph shows massive putrid empyema in the right pleural space. Culture yielded *Fusobacterium* sp, *Peptostreptococcus* sp, and microaerophilic streptococci.

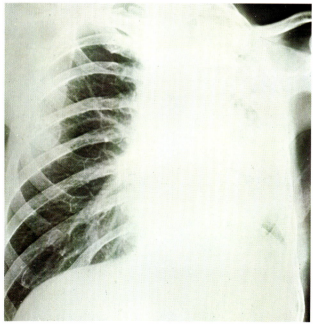

Figure 30. Pulmonary actinomycosis. Radiograph shows pulmonary actinomycosis involving entire left lung, with multiple abscesses and considerable loss of lung volume.

Figure 32. Janeway lesions, indicative of bacterial endocarditis.

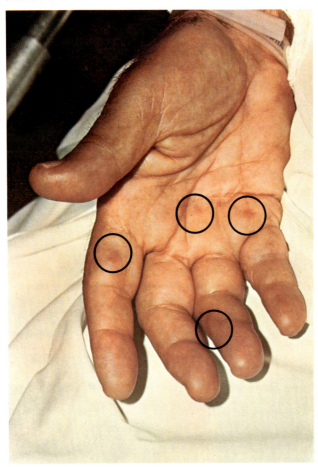

number of facultative organisms in the bowel flora, permitting *Bacteroides* to proliferate.

Speciation of isolates is important because, as stated previously, *B fragilis* is resistant to a number of antimicrobial agents, and other anaerobes may be resistant as well. Thorough surgical drainage and debridement of necrotic tissue are extremely important. The mortality resulting from *Bacteroides* bacteremia is approximately 30%.

BACTERIAL ENDOCARDITIS

In the past, anaerobes have seldom been reported as a cause of bacterial endocarditis. In large series of cases of endocarditis, only 1% to 2% of positive blood cultures yielded anaerobes; however, as many as 15% of the blood cultures were sterile, even in the presence of clinical or pathologic evidence strongly indicating endocarditis. It is likely that many of these so-called sterile cultures contained anaerobes that were not recovered because anaerobic techniques were inadequate. Therefore, the incidence of anaerobes as a cause of endocarditis is probably greater than 1% to 2%.

The signs and symptoms of endocarditis due to anaerobes are not significantly different from those of other types of endocarditis. Janeway lesions (Figure 32) can occur in both, for example. However, it has been reported that the incidence of embolization is higher and the incidence of preexisting heart disease is lower in endocarditis caused by anaerobes than in that caused by facultative organisms.

The main portals of entry for anaerobes that infect the heart valves are the mouth (especially with poor oral hygiene, periodontal disease, and tooth extraction), the gastrointestinal tract, and, to a lesser extent, the genitourinary tract. Identification of the portal of entry could indicate whether an anaerobe is the likely etiologic agent in bacterial endocarditis.

The most commonly isolated organisms in anaerobic bacterial endocarditis are *B fragilis* and species of *Fuso-*

bacterium, Clostridium, and Peptostreptococcus. Figure 33A shows Fusobacterium necrophorum growing in the bottom of a blood-culture bottle. A Gram stain of this growth shows the typical microscopic morphology of this organism (Figure 33B).

Appropriate antimicrobial therapy is of utmost importance in improving the prognosis of anaerobic endocarditis—a disease with a very high mortality. Specific identification of the organisms and antimicrobial dilution susceptibility tests with bactericidal endpoints are especially important. Previously, optimum therapy was hampered by the lack of a drug with bactericidal activity against B fragilis. Now, however, we have metronidazole, which shows consistently good bactericidal activity in vitro against B fragilis and has been used successfully in a few cases of B fragilis endocarditis. High-dose penicillin therapy may be useful against anaerobes other than B fragilis.

LIVER ABSCESS

The important role of anaerobic bacteria in the etiology of pyogenic liver abscess has not been fully recognized. A number of clinicians still believe that aerobic or facultative bacteria are the predominant cause of bacterial liver abscess; however, studies at our institutions indicate that over half of such abscesses are caused by anaerobic bacteria, despite failure to routinely obtain adequate anaerobic cultures in all cases.

Also, critical review of published series of pyogenic liver abscess suggests that anaerobes play an important role. It should be noted that the absence of anaerobes on culture does not necessarily mean they were not involved in the abscess. In some studies, either anaerobic cultures were not performed, anaerobic transport and culture techniques were inadequate, or no information was given. Furthermore, certain organisms such as coagulase-negative staphylococci, micrococci, and diphtheroids (which are likely contaminants)

Figures 33A and 33B. *Fusobacterium necrophorum.*

Figure 33A. Colonies of *Fusobacterium necrophorum* growing at the bottom of an anaerobic blood-culture bottle.

A

B

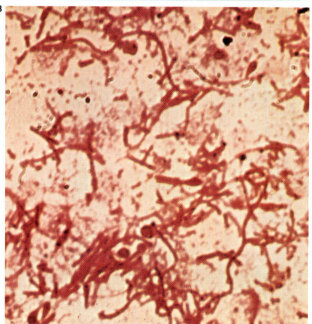

Figure 33B. Gram-stained smear of *Fusobacterium necrophorum* colonies from blood-culture bottle.

were considered to be significant aerobic isolates, as were certain other organisms that were present in very small numbers.

Liver abscess can result from extension of an adjacent or nearby infectious focus or can be related to other processes that permit invasion of tissues by an indigenous flora. Important underlying conditions include infected surgical wounds, inflammatory or malignant bowel disease (particularly appendicitis or diverticulitis), intestinal perforation, perirectal abscess, and cholecystitis. Pylephlebitis and cholangitis are also common forerunners of liver abscess. Embolic liver abscess, frequently multiple, may originate by way of the hepatic artery from foci anywhere in the body.

The clinical features of anaerobic liver abscess are indistinguishable from those of liver abscess in general. Solitary or multiple abscesses may be present; most are in the right lobe of the liver, in a posterior location. Amebic liver abscess may also be present in these locations. Epidemiologic history and parasitologic and serologic tests may help to distinguish this entity from pyogenic liver abscess.

A specific diagnosis can be made preoperatively in many cases by using good anaerobic blood culture techniques in conjunction with the usual procedures, such as radiographs, liver scans, and ultrasound, for diagnosing liver abscess. Radiographs in Figures 34-37 illustrate various characteristics of liver abscess. Liver photoscans following intravenous injection of radioisotope-labeled material (Figures 38-40) are particularly useful for making a diagnosis. Also, percutaneous aspiration of an abscess, preferably with ultrasound guidance, may be performed for diagnostic purposes.

Anaerobic and microaerophilic streptococci (Figure 41), *Fusobacterium necrophorum*, *F nucleatum*, *Clostridium* sp, *Bacteroides fragilis*, and *Actinomyces* sp are the anaerobes found most commonly in liver abscesses.

Figure 34. Elevated right diaphragm with right pleural effusion, both secondary to liver abscess.

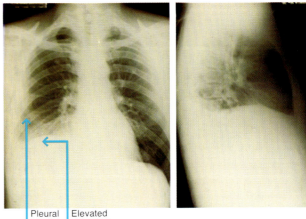

Pleural effusion | Elevated right diaphragm

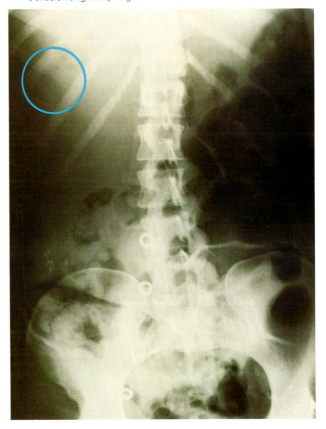

Figure 35. Multiple pockets of gas in a liver abscess (between 10th and 11th posterior ribs on right). This feature is very suggestive of anaerobic infection but is not specific for it.

Figure 36. Liver abscess. Celiac angiogram shows space-occupying lesion in the liver (note straightening of some vessels, with increased space between branches). Nephrogram is also apparent.

Figures 38A and 38B. Liver abscess. AP (anteroposterior) and lateral liver scans show location – superiorly in right lobe – and extent of abscess, which was caused by microaerophilic streptococci.

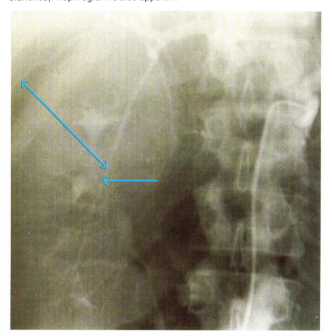

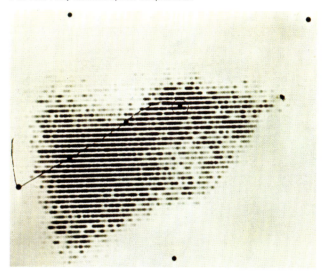

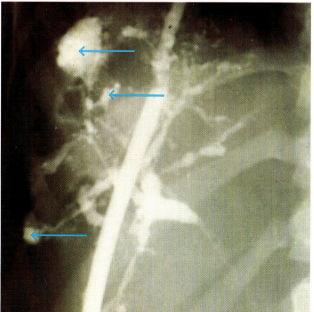

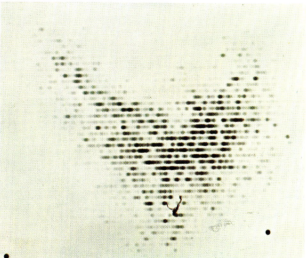

Figure 38B. Extent of abscess.

Figure 37. Contrast material injected through a draining abdominal-wall sinus in a young boy reveals multiple liver abscesses. (The arrows point to some areas of contrast material.)

Figures 39A and 39B. Liver abscess.

Figure 39A. AP liver scan of 13-year-old girl who had fever of undetermined origin and marked weight loss following a ruptured appendix, which was treated surgically. Blood cultures yielded *Bacteroides fragilis* in small numbers on two occasions, but scan shows no abnormalities.

Figure 40. Multiple liver abscesses. AP liver scan reveals many defects representing multiple liver abscesses.

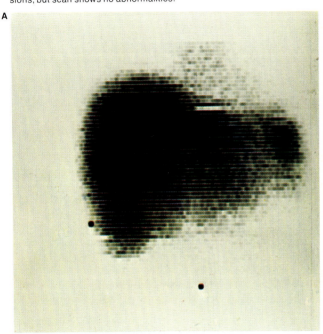

A

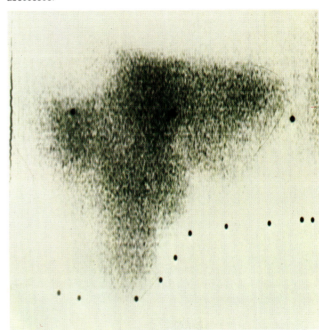

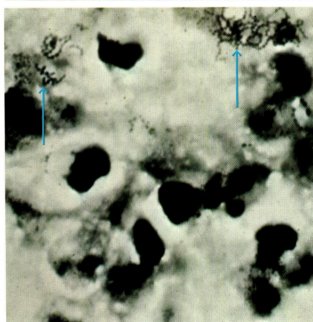

B

Figure 39B. Lateral scan of same patient reveals posteriorly located liver abscess.

Figure 41. Direct Gram stain of pus specimen from liver abscess reveals many chains of tiny streptococci; these proved to be microaerophilic streptococci on culture.

Surgical intervention is necessary, often immediately, for definitive therapy; however, antimicrobial therapy is certainly an important adjunct. In the absence of specific knowledge concerning the infecting organism, chloramphenicol represents a good choice for therapy. For abscesses in which the etiologic agent is believed to be an anaerobe but is not yet specifically identified, metronidazole or clindamycin are good choices. Metronidazole is also effective in treating amebic liver abscess.

As in many other types of anaerobic infections, therapy must be prolonged to minimize the possibility of complication or relapse. Patients who have a solitary abscess usually should be treated for at least two months, whereas patients who have multiple abscesses require therapy for at least four months. When surgical drainage of multiple abscesses is not possible, administration of antimicrobials by way of the umbilical vein may prove effective. Percutaneous aspiration of multiple abscesses, using ultrasound guidance together with antimicrobial therapy, may also prove successful in patients whose multiple abscesses cannot be drained surgically. Medical treatment, with or without percutaneous aspiration, has proved effective in a number of patients and could be considered in a patient who is not a candidate for surgery.

The prognosis is relatively good for patients whose abscesses are diagnosed and treated appropriately; however, patients who have multiple abscesses have a poorer prognosis than do those who have a solitary abscess.

BILIARY TRACT INFECTION

Anaerobes are not commonly involved in cholecystitis, but clostridia (and, rarely, nonsporulating anaerobes) may be involved in acute gaseous cholecystitis, a disease seen primarily in elderly diabetic patients. Gas forms along the contour of the gallbladder wall and tends to diffuse into surrounding tissues. A gas-fluid level may be noted within the gallbladder. Immediate surgery and penicillin therapy (10 to 20 million units IV daily) are recommended.

Biliary tract infections in the elderly, especially patients who have undergone multiple surgical procedures on the biliary tract, may involve *B fragilis*. These infections can in turn lead to bacteremia, septic shock, and death (Figure 42).

38

Figure 42. Summary of clinical course of a diabetic patient who had undergone a cholecystectomy. Complicating subhepatic abscess (mixed infection with *C perfringens* and facultative bacteria) developed, followed by *Bacteroides fragilis* bacteremia, septic shock, and death.

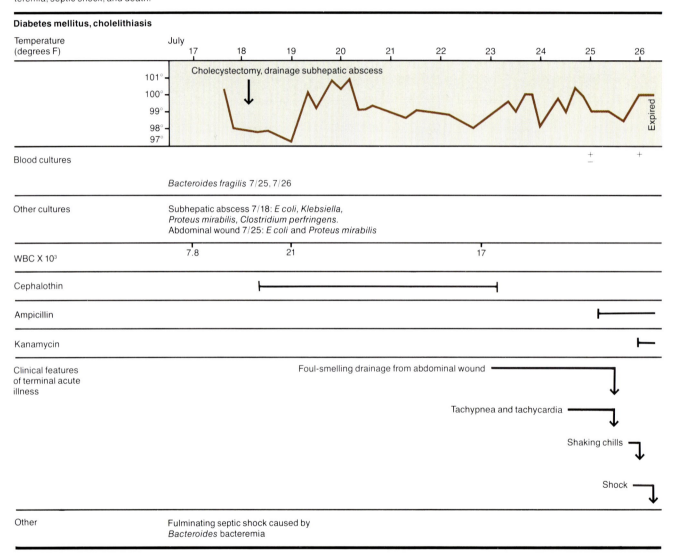

Diabetes mellitus, cholelithiasis

Temperature (degrees F)

July
17 18 19 20 21 22 23 24 25 26

101°
100°
99°
98°
97°

Cholecystectomy, drainage subhepatic abscess

Expired

Blood cultures
+ +
−

Bacteroides fragilis 7/25, 7/26

Other cultures
Subhepatic abscess 7/18: *E coli, Klebsiella, Proteus mirabilis, Clostridium perfringens.*
Abdominal wound 7/25: *E coli* and *Proteus mirabilis*

WBC X 10³
7.8   21   17

Cephalothin

Ampicillin

Kanamycin

Clinical features of terminal acute illness
Foul-smelling drainage from abdominal wound

Tachypnea and tachycardia

Shaking chills

Shock

Other
Fulminating septic shock caused by *Bacteroides* bacteremia

Figure 43. Gram stain of a specimen from an abdominal abscess. A variety of gram-positive and gram-negative cocci and bacilli are present.

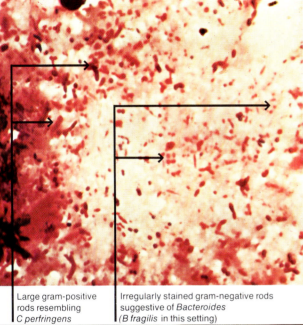

| Large gram-positive rods resembling *C perfringens* | Irregularly stained gram-negative rods suggestive of *Bacteroides* (*B fragilis* in this setting) |

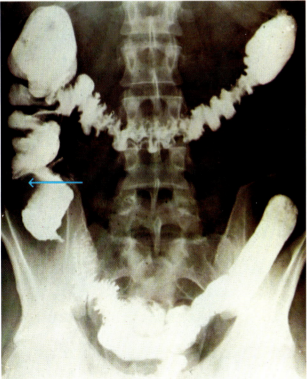

Figure 44. Appendiceal abscess. The arrow points to an area of extrinsic compression of the cecum by the abscess.

OTHER INTRA-ABDOMINAL INFECTIONS

Anaerobes are the prime cause of intra-abdominal abscesses and peritonitis. In fact, these organisms should be recovered in significant numbers from virtually every case of appendicitis with abscess formation and of peritonitis if proper transport and culture techniques are used. Because anaerobes make up more than 99% of the normal fecal flora, there is ample opportunity for these organisms to contaminate the peritoneum and extraperitoneal spaces. *Bacteroides fragilis*, other species of *Bacteroides*, and species of *Clostridium*, *Peptococcus*, and *Peptostreptococcus* are the most frequently isolated anaerobes. Commonly, a mixture of anaerobes and facultative organisms is recovered. Gram-stained smears of pus may show a variety of gram-positive and gram-negative cocci and bacilli with varying morphology (Figure 43).

These infections develop secondarily to problems such as appendicitis (Figure 44); diverticulitis with perforation (Figure 45); perforation associated with malignancy; bowel infarction or obstruction; and fecal contamination of a surgical wound. Figure 46 depicts small pockets of gas in an anaerobic subdiaphragmatic abscess that developed following a splenectomy. Although suggestive of anaerobic infection, gas in tissues is not specific for it because some facultative organisms, especially coliforms, are also gas producers.

Fever, leukocytosis, and pain and tenderness in the abdomen suggest the presence of an intra-abdominal abscess. A mass may be palpable on occasion. The development of board-like rigidity of the abdominal wall and rebound tenderness indicate that peritonitis has developed. Subphrenic abscesses may be very difficult to diagnose, and the patient may have an insidious and prolonged disease course.

Radiographic studies such as those shown in Figures 44-46 may be very helpful in making a correct diagnosis. Posteroanterior (PA) films of the abdominal area,

Figure 45. Diverticulitis with perforation and a localized peritonitis. The top arrow points to area where barium has extravasated through perforation into the peritoneal cavity; the other arrows point to diverticula of the colon.

Figure 46. Subdiaphragmatic abscess. Arrows point to gas within the abscess.

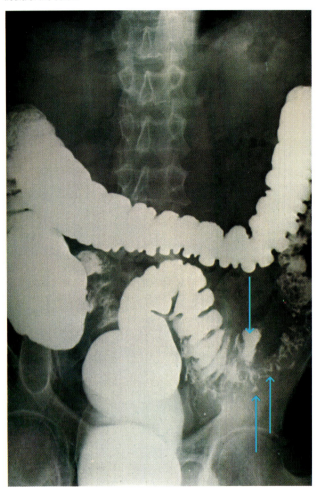

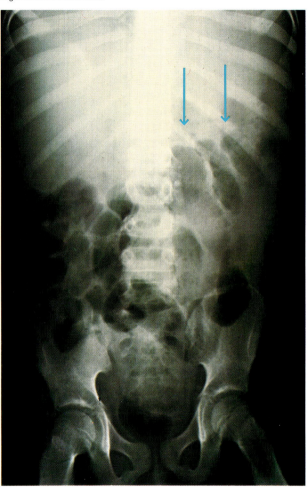

including the diaphragm, should be obtained to determine whether free air is present in the peritoneal cavity. The presence of free air indicates that perforation has taken place. Blood cultures, abdominal aspirate cultures, ultrasound, computerized tomography, and radioisotope scanning techniques are helpful in making a diagnosis. Blood cultures will frequently be positive for anaerobes (if cultured appropriately), and a four-quadrant-needle aspiration of the abdomen may reveal purulent material that can be cultured anaerobically as well as aerobically. Radioisotope scanning techniques can be very helpful in establishing a diagnosis of subphrenic abscess; a liver scan may show indentation of the surface of the liver (Figures 47A and 47B), and combined liver-lung scans may show an abnormally large abscess-containing space separating these organs (Figures 48-49). In general, combined liver-lung scans are less useful than gallium scans, computerized tomography, and ultrasound for diagnosis of subphrenic abscess, and of course these latter procedures are very useful for diagnosis of other intra-abdominal abscesses.

Abscesses require surgical drainage. Patients should be treated initially with drugs active against *B fragilis* as well as other anaerobes and facultative gram-negative rods. Chloramphenicol or a combination of either clindamycin or metronidazole and an aminoglycoside are frequently used. Cefoxitin may also be used but is not active against 5% to 10% of *B fragilis* strains and about 35% of clostridia other than *C perfringens*. The course of these infections is often prolonged and complicated. If multiple abscesses are present, they may be difficult to locate and may require repeated operations to ensure adequate drainage. Colostomy or ileostomy may also be required.

Serious complications include superinfection of wounds with antibiotic-resistant bacteria, bacteremia, pneumonia, fluid and electrolyte imbalance, and cardiac and renal disorders, especially in the elderly. These

Figures 47A and 47B. Radioisotope scan of liver shows indentation of the liver surface, caused by an abscess mass.

Figures 48A and 48B. Subdiaphragmatic abscess. Radioisotope scan of lung and liver shows abnormally wide space separating these two organs.

Figures 49A and 49B. Subdiaphragmatic abscess. Scan shows much greater separation between lung and liver than that seen in Figures 48A and 48B.

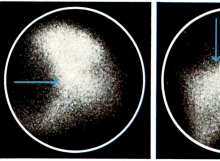

Figure 47A. Right lateral view.   **Figure 47B.** Anterior view.

Figure 48A. Anterior view.   **Figure 48B.** Right lateral view.

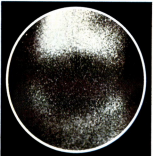

Figure 49A. Anterior view.   **Figure 49B.** Right lateral view.

42

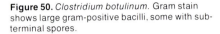

Figure 50. *Clostridium botulinum.* Gram stain shows large gram-positive bacilli, some with sub-terminal spores.

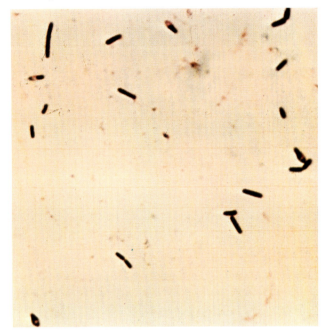

complications, combined with progressive weight loss, debilitation, and severe inanition, can lead to death. The introduction of intravenous hyperalimentation has significantly improved the nutrition of these patients, enabling them to better endure this long, difficult course.

The presently used preoperative oral bowel preparations include erythromycin or tetracycline in addition to oral neomycin or kanamycin. This regimen, used for a very limited period of time, minimizes the likelihood of overgrowth of anaerobes (and other resistant forms), which may be seen with the aminoglycosides used alone.

GASTROINTESTINAL INFECTIONS, INTOXICATIONS, AND OTHER DISEASE PROCESSES
Botulism

Clostridium botulinum, a common natural inhabitant of soil (including silt), may contaminate fruits, vegetables, fish, and other foods. If contaminated food that is to be canned or preserved is inadequately cooked and anaerobic conditions develop, *C botulinum* spores germinate; the resulting vegetative organisms then elaborate a potent neurotoxin. Then if the canned or preserved food is inadequately cooked before it is ingested, botulism develops. Although home-canned foods are more often the source of botulism, outbreaks have occurred after ingestion of commercially canned foods. Figure 50 shows a Gram stain of typical gram-positive bacilli with subterminal spores.

Once the toxin is ingested, it is absorbed in the stomach, then becomes fixed in cranial and peripheral nerves. The toxin blocks the transmission of nerve impulses in cholinergic nerve fibers by preventing the release of acetylcholine.

C botulinum is classified into types according to serologically distinct neurotoxins. Types A, B, and E are those usually responsible for food poisoning in humans. For example, type E botulism has been associated with the ingestion of preserved fish.

The salient clinical features of botulism are shown in Table 8. Usually, the first signs and symptoms occur within 18 to 36 hours (sometimes longer) after ingestion of contaminated food. Initially, nausea and vomiting may occur, then blurring of vision, double vision, and difficulty in swallowing. Motor paralysis may spread in a descending symmetrical pattern involving all the cranial nerves, the limbs, and even the respiratory muscles.

These clinical features, combined with a history of ingestion of canned foods (home or commercially prepared), are suggestive of botulism but are not adequate to make a firm diagnosis. Nor does recovery of C botulinum from the food or the patient, in itself, confirm the diagnosis. Absolute confirmation depends on the demonstration of the presence of botulinum toxin in the food or in the patient's serum. However, once botulism is even suspected, prophylactic measures such as gastrointestinal lavage, enema, and the administration of a trivalent antitoxin (after testing the patient for sensitivity to horse serum) should be instituted. The antitoxin contains antibodies for the three most common types of toxins (A, B, and E) encountered in humans.

Because circulating toxin has been detected in serum several weeks after contaminated food has been ingested, antitoxin should be given even if several days have elapsed since the patient ingested the food. Intensive supportive care, including assisted respiration, should be instituted as soon as symptoms develop. Mortality associated with botulism is still 25% to 35%.

Recently, "infant botulism" has been recognized as a distinct entity in which the ingested organism produces toxin in the small bowel. Honey is one known source of the organism. Because the disease is milder than conventional botulism and occurs in infants, diagnosis can be difficult.

Table 8.
Botulism

Incubation period
    Usually 18 to 36 hours, but can range from 2
    hours to 14 days

Gastrointestinal symptoms (appear first in one third of cases)
    Nausea, vomiting, abdominal pain
    Diarrhea early, then obstipation
    without pain

CNS involvement
    All cranial nerves except I and II
    may be involved
    Cranial nerves most commonly affected are
    III, IV, VI, IX, and X
    May involve innervation of limbs,
    intercostals, and diaphragm

Therapy
    Lavage, enema, cathartics
    Trivalent antitoxin

Prophylaxis
    Treat others exposed but not yet ill

Figure 51. Gram-stained smear of stool from a patient with *C difficile* pseudomembranous colitis. Typical large gram-positive bacilli, long and with parallel sides, are more numerous here than in normal stool. A few white blood cells are also present.

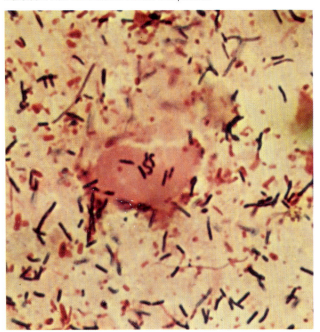

C perfringens Food Poisoning

C perfringens food poisoning is caused by ingestion of improperly cooked food that is contaminated with *C perfringens*. Severe cramping and diarrhea can result. The disease is self-limited, typically mild and of short duration, and usually requires only supportive therapy.

Enteritis Necroticans

Enteritis necroticans is associated with ingestion of foods heavily contaminated with type C *C perfringens*. Abdominal pain, vomiting, and diarrhea develop acutely, and shock and death may soon follow. Pathologically, widespread necrosis of the bowel (primarily small bowel) occurs. Studies indicate that a toxoid is effective in active immunization of humans.

Pseudomembranous Colitis

Clostridium difficile is a major cause of pseudomembranous colitis associated with antimicrobial therapy. Large numbers of causative organisms are found in the stools of patients with this disease. A Gram-stained stool smear (Figure 51) shows numerous gram-positive bacilli and a few white blood cells. In a normal stool smear, relatively few large gram-positive bacilli would be present. A toxin produced by *C difficile* causes a distinct cytopathic effect in many tissue culture cell lines. Another toxin, an enterotoxin, is probably primarily responsible for the pathogenic effects. There may be one or more additional toxins. Oral vancomycin is effective therapeutically, but relapses are not uncommon.

Intestinal Overgrowth of Bacteria

Intestinal overgrowth of bacteria can result from gastrointestinal surgical procedures or intestinal pathology. The "blind loop" syndrome is characterized by bacterial overgrowth in the upper portion of the small intestine. This syndrome can be caused by anaerobic bacteria that deconjugate bile acids, resulting in free bile acids that do not permit proper micelle formation and fat absorption. Vitamin B$_{12}$ malabsorption also results, probably from adsorption of the vitamin to bacteria.

Figure 52 shows the results of diagnostic tests and therapeutic measures in a patient with an afferent blind loop syndrome that developed in relation to previous gastric surgery; this patient also had multiple jejunal diverticula. Culture of a specimen from the afferent loop yielded anaerobes – *Bacteroides fragilis*, another *Bacteroides* sp, and a *Fusobacterium* sp – and facultative organisms. This patient's fecal fat level was high, and malabsorption of vitamin B$_{12}$, as measured by the Schilling test, was evident.

The first antibiotic used, oral neomycin, modestly reduced the number of facultative organisms but had no effect on the anaerobes, the fecal fat level, or vitamin B$_{12}$ malabsorption. The second antibiotic used was active against anaerobes and eliminated them without affecting the facultative organisms. Fecal fat level and vitamin B$_{12}$ absorption returned to normal.

FEMALE GENITAL TRACT INFECTION
A large variety of female genital tract infections involve anaerobic bacteria (Table 9). Conditions predisposing to or associated with anaerobic infections of the female genital tract include pregnancy; the puerperium (particularly when there is premature rupture of the membranes, prolonged labor, or postpartum hemorrhage); abortion (spontaneous or induced); malignancy; irradiation; obstetric or gynecologic surgery; cervical cauterization; endocervical or vaginal stenosis; uterine fibroids; disrupted architecture resulting from old gonococcal salpingitis; and the use of intrauterine contraceptive devices (IUDs).

The most dramatic of these infections occurs when clostridia from the normal vaginal flora or from outside the vagina invade the uterus. The first stage of such an infection, which typically occurs following criminal

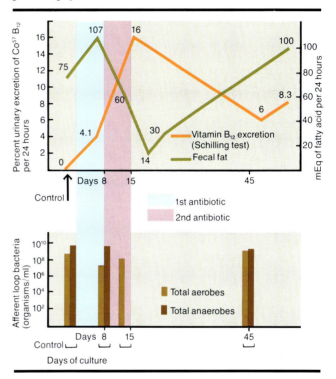

Figure 52. Hospital course of a patient with afferent "blind loop" syndrome that developed following gastric surgery.

Table 9.
Female genital tract infections involving anaerobes

Endometritis, pyometra

Myometritis

Parametritis

Pelvic cellulitis, pelvic abscess

Pelvic thrombophlebitis

Peritonitis, intra-abdominal abscess

Bacteremia, with or without
shock or intravascular hemolysis

Metastatic infection –
particularly from lung, liver, brain,
heart valves, kidneys

Wound infection or abscess

Vulvar abscess

Bartholinitis, Bartholin abscess

Skenitis, Skene's abscess

Paraclitoroidal abscess

Vaginitis, abscess of vaginal wall,
paravaginal abscess

Salpingitis, tubal abscess

Ovarian or tubo-ovarian abscess

Abscesses of adjacent parts –
groin, perirectal, ischiorectal, or
paraurethral areas, leg, abdominal wall

Chorioamnionitis

Fetal emphysema

Intrauterine or neonatal pneumonia,
sepsis

abortion, is fetal emphysema, a localized chorioam-nionitis with bacterial invasion of the dead fetus and the placenta. Although this is a relatively benign process, it can be difficult to distinguish from more serious infections because of the presence of a gaseous vaginal discharge, crepitus of the uterine wall, and striking radiographic findings. Only fetal cultures yield clostridia. Treatment requires evacuating the uterus and instituting antimicrobial therapy.

The next stage of infection is a low-grade endometritis that produces vaginal discharge and tenderness of the uterus. There is no toxemia, and there may or may not be gas formation. In this stage, clostridia can be found on culture of the uterine discharge, but care must be taken to avoid contaminating the specimen with normal vaginal or cervical flora. Curettage and penicillin constitute appropriate therapy.

The most severe stage of this infection is endometritis and myometritis with necrosis, perforation, peritonitis, and sepsis. The patient exhibits significant toxicity, fever, tachycardia disproportionate to the fever, uterine and abdominal pain, and a foul gaseous vaginal discharge that reveals numerous clostridia on Gram stain. Blood culture is frequently positive for anaerobes. Patients with significant sepsis manifest intravascular hemolysis, jaundice, hemoglobinemia and hemoglobinuria, shock, and renal shutdown. Diffuse hemorrhage may occur. Mortality in this stage is 50% to 75%. Appropriate treatment includes surgical removal of necrotic tissue (often necessitating hysterectomy), large doses of penicillin administered intravenously (assuming that the infection does not involve penicillin-resistant organisms such as *B fragilis* and *E coli*), and general supportive care.

Another infection caused by anaerobes is pelvic actinomycosis, which is primarily a chronic infection. Although it is uncommon, there have been a number of reports of infection in women using IUDs.

Although clostridial infections are the most impressive and most serious of female genital tract infections, those of all types listed in Table 9 are much more commonly produced by anaerobic and microaerophilic cocci, *Bacteroides* (including *B fragilis*), and other organisms listed in Table 10.

URINARY TRACT INFECTION

A number of noncritical reports on the subject of anaerobic urinary tract infections have caused confusion because they failed to consider the presence of anaerobes as normal flora on the urethral mucosa; however, anaerobes may cause as many as 1% to 2% of infections involving the urinary system. The various types of documented anaerobic genitourinary tract infections are listed in Table 11.

Anaerobic infections of the urinary tract can result from bacteremia; migration of organisms from the bowel; the indigenous urethral flora (particularly after instrumentation); the introduction of bacteria during surgery; malignancy; obstruction; and previous renal tuberculosis. Kidney infections may be associated with stones, particularly tissue necrosis adjacent to a stone.

Distinctive features that may be seen in anaerobic infection of the urinary tract, as in many other anaerobic infections, include tissue necrosis, gas in tissues, and foul odor.

In obtaining specimens for culture, it is important to avoid contaminating the specimen with the normal urethral flora. Uncontaminated urine specimens can be obtained by percutaneous suprapubic bladder aspiration (Figure 53). Treatment includes debridement, drainage, and appropriate antimicrobial therapy.

SKIN, SOFT-TISSUE,
AND MUSCLE INFECTIONS
AND INTOXICATIONS

Anaerobic infections of skin and soft tissue may de-

Table 10.
Organisms involved in female genital tract infections

Anaerobes

Peptostreptococcus
Peptococcus
Veillonella
Bacteroides bivius
Bacteroides disiens
Bacteroides fragilis
Bacteroides melaninogenicus
Other *Bacteroides* sp
Fusobacterium necrophorum
Other *Fusobacterium* sp
Clostridium perfringens
Actinomyces israelii

Other organisms

Escherichia coli
Other Enterobacteriaceae
Microaerophilic gram-positive cocci
Group A beta-hemolytic streptococci
Enterococci
Other streptococci
Staphylococcus aureus
Gonococcus
Gardnerella vaginalis
Proteus
Pseudomonas

48

Table 11.
Urinary tract infections involving anaerobes

Periurethral cellulitis, abscess, or gangrene

Cowperitis

Chronic prostatitis

Prostatic abscess

Cystitis

Pyelocystitis

Pyelonephritis

Pyonephrosis

Renal abscess

Perirenal abscess

Testicular abscess

Wound infection following nephrectomy,
nephrostomy, or other open surgery

Gas gangrene involving various structures
including bladder, kidneys, scrotum,
and penis

Actinomycosis – usually a chronic,
circumscribed renal parenchymal lesion;
may progress to pyonephrosis or
perinephric abscess.

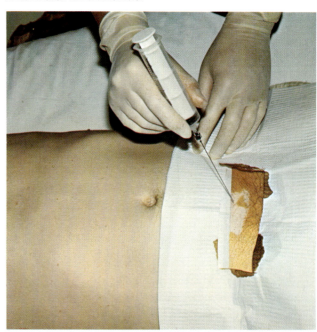

Figure 53. Collection of urine specimen by supra-pubic bladder aspiration. This technique avoids contamination with urethral flora.

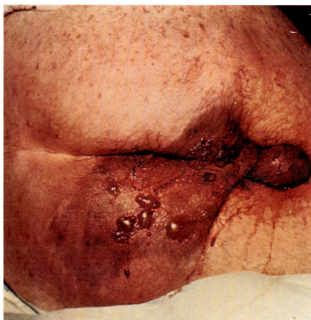

Figure 54. Buttocks and perineal region of a patient with gas gangrene. Prominent features are extensive ecchymoses and necrosis with multiple bullae.

velop in areas that are traumatized or devitalized as a result of surgical wounds and injuries, including bites. They may also develop as a result of ischemia in the extremities of patients with atherosclerosis, diabetes mellitus, or both. Different causative organisms are likely to be involved in different locations and types of injuries. For example, infections involving the lower half of the body are most likely to be contaminated with fecal flora, such as *Clostridium* sp, *B fragilis*, *Peptostreptococcus* sp, and others. Bites, however, are likely to be contaminated with mouth flora, such as *B melaninogenicus*, *Fusobacterium* sp, anaerobic cocci, and others.

Clostridial Myonecrosis (Gas Gangrene)

Perhaps the most severe infection caused by anaerobes is clostridial myonecrosis, or "gas gangrene." The causative organisms are species of clostridia, including *C perfringens*, *C histolyticum*, *C septicum*, *C novyi*, and *C bifermentans*. They are gram-positive sporeforming bacilli that can be found in soil and are part of the normal human bowel flora. Consequently, soil or fecal contamination of an injury or fecal contamination of a surgical wound can lead to gas gangrene. The buttocks, thighs, and perineum are common sites for this infection. It can occur following septic abortion and in the presence of ischemic vascular disease.

Some of the typical clinical features of gas gangrene are illustrated in Figures 54-57. In Figure 54, wide areas of ecchymosis, necrosis, and edematous skin are present. A dark-red serous material exudes from the wound, and numerous gas-filled vesicles and bullae are present. Figure 55 shows tense, shiny, pale edema of the skin and patchy areas of necrosis. Figure 56 shows radiographic features of gas in the soft tissues of the arm of a man with a compound fracture of the humerus. This infection developed as a result of an injury sustained in a motorcycle accident. Figure 57 illustrates a clostridial wound infection that developed following

Figure 55. Gas gangrene developing after a compound fracture of the tibia and fibula. Note the necrotic skin and prominent bullae.

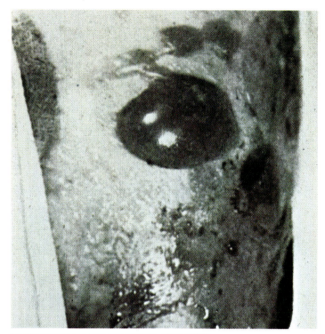

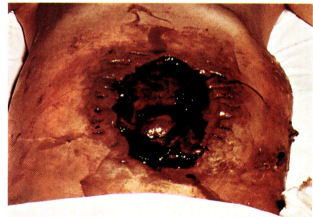

Figure 56. Radiograph showing gas shadows in the area of a compound fracture of the humerus where gas gangrene has developed.

Figure 57. Gas gangrene of the abdominal wall following bowel surgery.

Figure 58. Histologic section from a gas gangrene infection showing myonecrosis and gas in tissues.

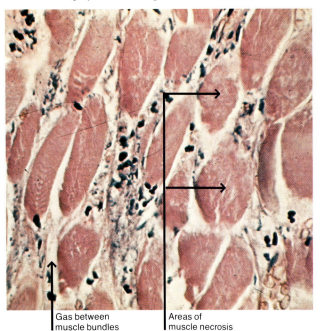

Gas between muscle bundles | Areas of muscle necrosis

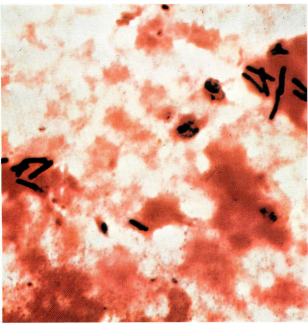

Figure 59. Gram-stained smear of exudate from gas gangrene (Figure 57) showing large gram-positive bacilli characteristic of *Clostridium perfringens*.

bowel resection for adenocarcinoma; note the bronze discoloration of the skin. Fecal contamination of the wound led to widespread necrosis of the abdominal wall. This patient survived after treatment with extensive debridement, penicillin, antitoxin, and hyperbaric oxygen (although the role of the latter two therapeutic modalities must be considered uncertain).

Additional prominent features of gas gangrene are pain that develops early and toxemia that develops somewhat later. Figure 58 illustrates a muscle biopsy with necrosis of tissue and abundant gas in the tissues.

Bacteriologic features of clostridia are shown in Figures 59-61. Figure 59 shows a Gram-stained smear of the exudate from the infection shown in Figure 57: numerous large gram-positive rods are present, although no spores are evident. The development of double zones of hemolysis when *C perfringens* is grown on blood agar plates (Figure 60) and the Nagler reaction (Figure 61) are also helpful in the laboratory identification of *C perfringens*. In the Nagler reaction, the clostridial lecithinase attacks the egg yolk in the medium and causes an opaqueness about the area of bacterial growth, as shown on the right side of the plate. The left side was treated with *C perfringens* type A antitoxin (an antilecithinase), thus preventing opacification.

Gas gangrene is a clinical diagnosis. The existence of typical organisms – even *C perfringens* – on a Gram-stained smear or on culture does not establish the diagnosis, because the organism can be present without causing disease. The typical clinical picture must also be present.

C septicum can cause rapidly progressive myonecrosis. *C septicum* infection is found more often in patients with underlying malignancy, particularly carcinoma of the cecum.

Sepsis caused by species of *Clostridium* may occur secondary to gas gangrene or other types of infections, and it can be a serious complication. The alpha toxin of *C perfringens* may cause shock and massive hemolysis, which often leads to renal failure.

The prognosis for patients with gas gangrene is poor. Adjunctive measures such as therapy with penicillin and hyperbaric oxygen may be useful, but thorough debridement is the key to successful management. This may require extensive and even radical surgery (including amputation).

Several infections, some of which are caused by *Clostridium* sp and some by other organisms, exhibit certain clinical features suggestive of gas gangrene. Examples are synergistic nonclostridial anaerobic myonecrosis (also known as necrotizing cutaneous myositis or synergistic necrotizing cellulitis), anaerobic streptococcal myositis, infected vascular gangrene, necrotizing fasciitis, and anaerobic cellulitis. Table 12 lists the differentiating characteristics of each of these infections. Each will also be discussed.

Synergistic Nonclostridial Anaerobic Myonecrosis

Synergistic nonclostridial anaerobic myonecrosis has characteristics similar to those of gas gangrene (Table 12). It is a virulent soft-tissue infection involving skin, subcutaneous tissue, fascia, and muscle. Extensive confluent necrotic liquefaction ("dishwater pus") may be present. *Bacteroides* sp, anaerobic streptococci, or both are found, together with aerobic or facultative gram-negative bacilli, in these infections.

Anaerobic Streptococcal Myositis

Anaerobic streptococcal myositis is characterized by edema and a seropurulent exudate. Severe pain and toxemia develop later in the course of this illness than in clostridial myonecrosis, and with proper therapy the prognosis is good. Debridement plus antibiotic therapy is often sufficient.

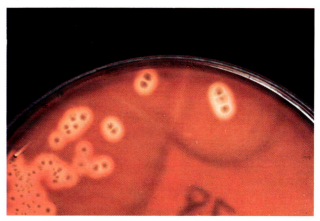

Figure 60. Colonies of *C perfringens* when grown on blood agar plates often produce a small inner zone of complete hemolysis and a wide outer zone of partial hemolysis.

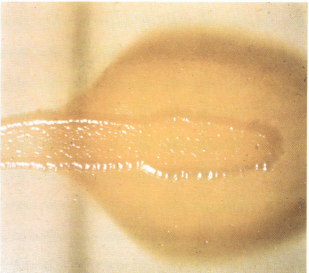

Figure 61. Identification of clostridia by the Nagler reaction. The organism has been streaked horizontally across the plate. When *C perfringens* is grown on egg-yolk agar, the action of lecithinase causes opacification of the surrounding medium, except in areas where the activity is inhibited by prior swabbing with antitoxin to alpha toxin (left side of plate).

Table 12.
Differentiating characteristics of certain similar anaerobic infections of skin and soft tissue

	Clostridial myonecrosis (gas gangrene)	Synergistic non-clostridial anaer-obic myonecrosis	Anaerobic streptococcal myositis	Infected vascular gangrene	Necrotizing fasciitis	Anaerobic cellulitis
Incubation	Usually less than 3 days	Variable, 3-14 days	3-4 days	More than 5 days, usually longer	1-4 days	Almost always more than 3 days
Onset	Acute	Acute	Subacute or insidious	Gradual	Acute	Gradual
Toxemia	Very severe	Marked	Severe only after some time	Nil or minimal	Moderate to marked	Nil or slight
Pain	Severe	Severe	Variable, as a rule fairly severe	Variable	Moderate to severe	Absent
Swelling	Marked	Moderate	Marked	Often marked	Marked	Nil or slight
Skin	Tense, often very pale	Minimal change	Tense, often with a coppery tinge	Discolored, often black & desiccated	Pale-red cellulitis	Little change
Exudate	Variable, may be profuse: serous & blood stained	"Dishwater" pus	Very profuse, seropurulent	Nil	Serosanguinous	Nil or slight
Gas	Rarely pro-nounced except terminally	Not pronounced; present in 25% of cases	Very slight	Abundant	Usually not present	Abundant
Smell	Variable, may be slight, often sweetish	Foul	Very slight, often sour	Foul	Foul	Foul
Muscle	Marked change	Marked change	At first little change except edema	Dead	Viable	No change

Facultative bacteria may also produce some of these infections.

Figure 62. Infected vascular gangrene caused by *Bacteroides fragilis* in an amputation stump. Abundant gas is present in the tissues.

Figure 63A. Anaerobic cellulitis of the foot, caused by infection with *Bacteroides fragilis* and facultative organisms. Extensive subcutaneous gas is present. The patient was diabetic.

Infected Vascular Gangrene

Infected vascular gangrene is primarily a problem of poor blood supply with infection engrafted on vascular ("dry") gangrene. The infection is characterized by foul-smelling discharge, gas in tissues, and marked swelling, but there is relatively little systemic toxemia. The infection can be caused by a variety of anaerobes, including *C perfringens* and *B fragilis*, and facultative organisms.

Figure 62 shows radiographic features of *B fragilis*-infected vascular gangrene in an amputation stump; abundant gas is present in the tissues. In this type of gangrene, ischemia and infection resulting from fecal contamination remain a problem, even after removal of the originally infected area.

Necrotizing Fasciitis

Necrotizing fasciitis is a serious infection that spreads rapidly along fascial planes and is commonly caused by *Staphylococcus aureus* or *Streptococcus pyogenes*. Anaerobes—especially clostridia and bacteroides—can also be the cause.

Anaerobic Cellulitis

Anaerobic cellulitis, as defined by MacLennan, is a localized infection of soft tissue that does not involve muscle. The infection is characterized by foul-smelling discharge and gas in tissues. Pain, swelling, and exudate may be slight or absent, and there is little or no systemic toxicity; however, the infection can spread rapidly and become extensive. Clostridia, gas-producing nonsporeforming anaerobes, and facultative organisms can produce this infection, but an infection that involves clostridia is likely to be more severe.

Figures 63A and 63B show features of anaerobic cellulitis involving *B fragilis* and facultative organisms. Extensive subcutaneous gas is present in this large necrotic ulcer on the foot of a diabetic patient. Treatment

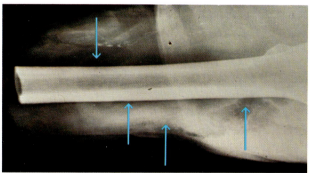
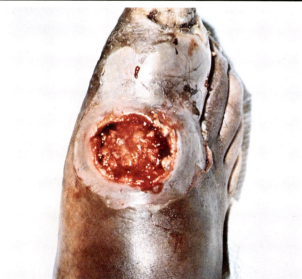
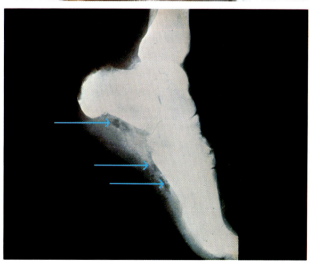

Figure 63B. Radiograph showing extensive gas in the tissues of the infected foot.

of this type of infection requires aggressive surgical debridement and antibacterial therapy. Osteomyelitis is a common sequela. Persistent or spreading infection and significant vascular disease often dictate the need for early amputation.

Progressive Bacterial Synergistic Gangrene

Progressive bacterial synergistic gangrene is a more unusual soft-tissue infection involving anaerobes. Figures 64A and 64B illustrate a small lesion that rapidly developed into synergistic gangrene, necessitating amputation of part of the index finger, despite administration of high-dose antibiotic therapy. *Peptostreptococcus* sp and *Staphylococcus aureus*, the two organisms most often involved in this disease process, were cultured from the wound. The organisms act synergistically to produce necrosis and ulceration of both skin and subcutaneous tissues.

Bite Wounds

Anaerobic infections associated with human or animal bites can be extremely virulent, resulting in progressive widespread tissue destruction. Anaerobic organisms commonly involved include *B melaninogenicus*, other bacteroides, *Fusobacterium* sp, and anaerobic cocci. Before the availability of antibiotics, bite-wound infections commonly led to a need for amputation. Today, such radical surgery is rarely necessary if appropriate high-dose antibiotic therapy (primarily penicillin) is promptly instituted.

Meleney's Ulcer (Chronic Undermining Ulcer)

Meleney's ulcer is another unusual soft-tissue infection. It is a slowly progressive, deep, subcutaneous infection that spreads diffusely, producing multiple sinuses and necrotic ulcers with undermining of the intervening skin. Such ulcers heal very slowly, even after adequate debridement and treatment with antibiotics. Micro-

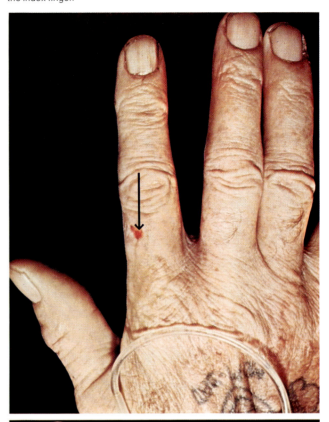

Figure 64A. Progressive bacterial synergistic gangrene. A rapidly developing necrotic infection caused by *Peptostreptococcus* and *Staphylococcus aureus*. Initial lesion was a small ulcer on the index finger.

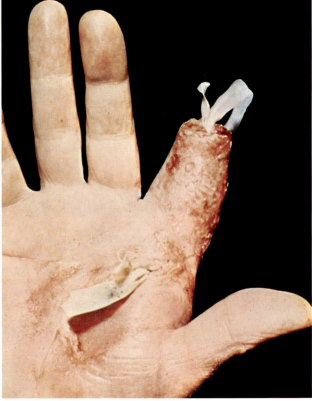

Figure 64B. With progression, amputation of the distal portion of the finger was necessary. A through-and-through drain extends from the amputation site and from the palm.

aerophilic streptococci are the causative organisms; however, anaerobic organisms are occasionally found on culture.

TETANUS

Clostridium tetani is ubiquitous in nature and can easily contaminate even minor wounds. As a result, tetanus can occur following conventional wounds of various types, and it may be seen in addicts who inject the drug subcutaneously. The organism grows locally in the wound but is not invasive and, in itself, is harmless; however, it elaborates a neurotoxin that acts on the spinal cord to reduce central inhibitory activity. As a result, excess motor activity and spasticity develop.

The early clinical features of tetanus are listed in Table 13. Frequently, the earliest sign is trismus, or "lockjaw." Pharyngeal muscle spasm results in dysphagia, and eventually generalized tonic convulsions develop. A Gram-stained smear of wound exudate may, on occasion, reveal typical clostridia with terminal spores (Figure 65), although the diagnosis of tetanus is made on clinical grounds.

Aggressive supportive care is necessary, with special emphasis on sedation and respiratory care. Furthermore, antitoxin and toxoid should be administered (each at a different site). Tetanus hyperimmune gamma globulin of human origin should be used to avoid the risks associated with the horse serum antitoxin. Debridement of the wound and administration of penicillin are also indicated.

With regard to prophylaxis of tetanus, people who have received a three-dose primary immunization series plus at least one booster dose every 10 to 15 years thereafter can be considered adequately immunized, except in cases of wounds associated with a high risk of tetanus. Therefore, routine boosters at the time of each injury are not usually necessary. Furthermore, too frequent booster doses may increase the likelihood of hypersensitivity reactions to the immunization.

Table 13.
Tetanus: Early signs and symptoms

Tension or cramps and twitching in muscles around wound
Increased reflexes in wounded extremity
Stiffness of jaw muscles; mild pains in facial muscles
Slight difficulty in swallowing
Stiffness of neck
Constipation
Headache, backache, general irritability and restlessness
Sweating, tachycardia
Anxious facies

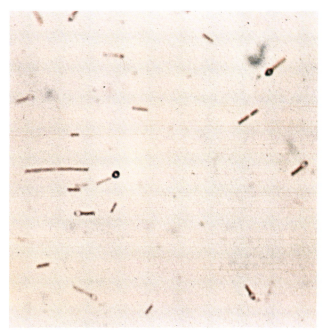

Figure 65. *Clostridium tetani.* Gram stain of culture shows large gram-positive bacilli, some with terminal spores.

Table 14.
Clues to the presence of anaerobes in osteomyelitis

Foul odor of discharge

Gas in soft tissues or in discharges

Tissue necrosis

Absence of potential pathogens on aerobic culture
(presence doesn't rule out concomitant
involvement of anaerobes)

Unique morphology of anaerobes on Gram stain

Diabetes

Neuropathy

Impaired circulation to the affected bone

Other underlying conditions suggestive of
anaerobic infection

Portals of entry suggesting anaerobes may be the
etiologic agent

BONE AND JOINT INFECTIONS

Osteomyelitis involves anaerobic bacteria much more commonly than has been recognized, but additional clinical and bacteriologic studies are needed to better document the significance of anaerobes in bone infections. A number of cases document osteomyelitis following sepsis caused by anaerobic bacteria, and actinomycosis involving bone is a well-established entity. Vertebral actinomycosis can be distinguished from tuberculous or pyogenic involvement of the spine; vertebral actinomycosis is characterized by the absence of vertebral collapse and the preservation of the disc space, whereas collapse occurs with tuberculous or pyogenic involvement. Actinomycotic infections and other types of osteomyelitis involving anaerobes may result from extension of soft-tissue infections or may be associated with trauma, surgery, or malignancy.

The role of anaerobes is not as well established in chronic forms of osteomyelitis as it is in acute blood-borne disease. In the preantibiotic era, infections of the bones of the jaw, skull (including the mastoids), and cervical spine frequently occurred secondary to pyorrhea, Vincent's angina, sinusitis, infections of the neck, and chronic otitis media. Such infections are less common now but are still significant problems. Although there is evidence that chronic osteomyelitis of long bones may involve anaerobes, it remains to be determined how often such infections are primary and how significant the anaerobes are in the case of open lesions where secondary infection or wound colonization is a possibility. If the same species of anaerobes are cultured repeatedly from such lesions, they are very likely significant.

Clues to the presence of anaerobes in osteomyelitis (but not necessarily unique to this infection) are listed in Table 14. Anaerobes most commonly isolated are not only the gram-negative bacilli (particularly the *Bacteroides melaninogenicus* and *B fragilis* groups) but also

Figure 66. Compound fracture of the leg with
Bacteroides infection complicated by pyarthrosis
of the hip. Draining sinuses are present.

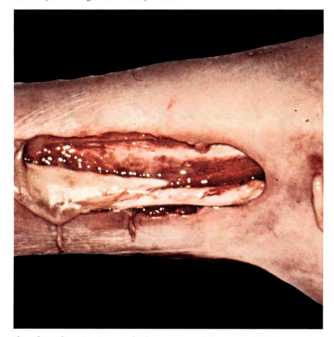

the fusobacteria and the anaerobic cocci. Figure 66
shows a compound fracture of the leg with *Bacteroides*
infection complicated by pyarthrosis of the hip.

In diabetic patients, anaerobes are relatively common in infections associated with diabetic gangrene
of the foot, with or without bone involvement. Anaerobes are also common in less severe infections, such as
trophic ulcers of the foot, again with or without bone
involvement.

There are a number of well-documented cases of
purulent arthritis caused by anaerobic bacteria. In those
cases, predisposing factors included poor host resistance, underlying joint disease, corticosteroid therapy,
and anaerobic infection elsewhere in the body. Any
joint may be affected, but the sternoclavicular and sacro-
iliac joints show peculiar susceptibility to anaerobic
infection. The majority of cases of anaerobic arthritis
are caused by gram-negative anaerobic bacilli and especially by *Fusobacterium necrophorum*.

Diagnosis and Bacteriology of Anaerobic Infections

ANAEROBES AS NORMAL FLORA

The major anaerobic infections and intoxications in humans are caused by anaerobic bacteria – or their toxins – that are indigenous to the normal flora, particularly bacteria of the mouth, intestinal tract, and female genital tract (Table 15). Occasionally such infections arise from exogenously acquired bacteria; soil is a common source of clostridia.

Anaerobic bacteria are usually more abundant as indigenous flora than are other microorganisms. For example, anaerobes outnumber aerobes at least 10 to 1 in the mouth and 1000 to 1 in the lower intestine.

The mouth and upper respiratory tract are inhabited by large numbers of various potentially pathogenic anaerobes. Aspiration pneumonia and lung abscess are frequently caused by nonsporeforming anaerobes such as *Bacteroides melaninogenicus*, *Fusobacterium* sp, and anaerobic cocci – all residents of the mouth and upper respiratory tract. Spirochetes also inhabit the mouth and are found in oral and pulmonary disease, but their significance is unknown.

The intestinal tract is inhabited by large numbers of almost every group of anaerobes known to cause infection in humans. Those most abundant are the *Bacteroides fragilis* group, anaerobic cocci, *Clostridium* sp, *Bifidobacterium* sp, and *Eubacterium* sp. The latter two genera are not major pathogens. Intestinal anaerobes, especially *B fragilis*, are almost always involved in infections that develop following bowel surgery.

A few studies show that the normal flora of the genitourinary tract, except for the vagina, consists primarily of anaerobic gram-negative nonsporulating bacilli. Anaerobic cocci are also commonly found. Although *Lactobacillus* sp is the predominant organism in the vagina, most vaginal infections are caused by the anaerobic cocci or *Bacteroides* sp.

The major groups of anaerobic bacteria indigenous to the skin are propionibacteria (anaerobic diphtheroids) and anaerobic gram-positive cocci. *Clostridium* sp may also be found but probably represent transient contamination from the soil or feces. Also, perineal and perianal skin may often be contaminated with intestinal flora.

The continued use of sophisticated methods of collection, cultivation, and identification of anaerobic bacteria will undoubtedly lead to an extension of our knowledge concerning the various genera and species indigenous to specific sites in the body.

CLUES TO THE PRESENCE OF ANAEROBIC INFECTIONS

Certain clinical signs, symptoms, and conditions are characteristic of anaerobic infections, as are certain bacteriologic features. Clinical and bacteriologic clues to the presence of anaerobic infections are listed in Tables 16 and 17, respectively.

COLLECTION OF SPECIMENS FOR DIAGNOSIS OF ANAEROBIC INFECTIONS

Proper procedure dictates that all clinical specimens destined for a complete bacteriologic analysis be cultivated anaerobically as well as aerobically. Collection of material that is normally sterile, such as blood and spinal fluid, requires only the usual precautions to decontaminate the skin before puncturing it to obtain the specimen. However, most anaerobes associated with infections are also present on mucous membranes or other areas of the body as part of the indigenous flora. Moreover, the anaerobes at these sites are frequently so numerous that if a clinical specimen comes into contact with only a minute portion of normal flora, misleading culture results will be obtained. Therefore, special collection procedures must be used to obtain specimens. ***Lung Abscess, Pneumonia, and Other Pulmonary Infections:*** Because of the large number of anaerobes present in saliva, sputum specimens are not suitable for anaerobic culture. Bronchoscopically obtained speci-

Table 15.
Presence of anaerobic bacteria as part of the normal flora in humans

Location	Clostridium	Nonsporulating bacilli							Cocci	
		Gram-positive					Gram-negative		Gram-positive	Gram-negative
		Actino-myces	Bifido-bacterium	Eubacte-rium	Lacto-bacillus	Propioni-bacterium	Bacte-roides	Fuso-bacterium		
Skin	0	0	0	±	0	2	0	0	1	0
Upper respiratory tract**	0	1	0	±	0	1	1	1	1	1
Mouth	±	1	1	1	1	±	2	2	2	2
Intestine	2	±	2	2	1	±	2	1	2	1
External genitalia	0	0	0	U	0	U	1	1	1	0
Urethra	±	0	0	U	±	0	1	1	±	U
Vagina	±	±	1	±	2	1	1	±	1	1

key to symbols:
U = unknown
0 = not found or rarely present
± = irregularly present
1 = usually present
2 = usually present in large numbers
† = includes anaerobic, microaerophilic, and facultative strains

**includes nasal passages, nasopharynx, oropharynx, and tonsils

Table 16.
Clinical clues to presence of anaerobic infection

Foul-smelling discharge

Location of infection in proximity to a mucosal surface

Necrotic tissue, gangrene, pseudomembrane formation

Gas in tissues or in discharges

Endocarditis with negative routine blood cultures

Infection associated with malignancy or other process that results in tissue destruction

Infection related to the use of aminoglycosides (oral, parenteral, or topical)

Septic thrombophlebitis

Infection following human or other bites

Black discoloration of blood-containing exudates; these exudates may fluoresce red under ultraviolet light (B melaninogenicus infections)

Presence of "sulfur granules" in discharges (actinomycosis)

Classic clinical features of gas gangrene, actinomycosis

Clinical conditions predisposing to anaerobic infection (septic abortion, gastrointestinal surgery, and other surgical procedures and diseases)

Table 17.
Bacteriologic clues to presence of anaerobic infection

Unique morphology on Gram stain

Failure to grow, aerobically, organisms seen on Gram stain of original exudate (Note: Failure to obtain growth in thioglycolate broth is not adequate assurance that anaerobes are not present)

Growth in anaerobic zone of fluid media or of agar deeps

Growth anaerobically on media containing 75 to 100 μg/ml of kanamycin, neomycin, or paromomycin (or a medium also containing 7.5 μg/ml of vancomycin in the case of gram-negative anaerobic bacilli)

Gas, foul odor in specimen or on culture

Characteristic colonies on agar plates when cultured anaerobically

Young colonies of B melaninogenicus may fluoresce red under ultraviolet light

mens may be suitable if obtained using special techniques, including the use of a double-lumen catheter with a distal polyethylene glycol plug, but this remains to be established. Transtracheal needle aspiration (shown in Figure 16) or direct lung puncture are the optimal methods of obtaining these specimens unless empyema fluid is available.

Other Abscesses: The unbroken skin or mucosal surface should be decontaminated and the pus removed with a syringe. This method is preferred to using a swab on a portion of an exposed lesion, because saprophytic anaerobes unrelated to the infection may be collected on the swab and cultured.

Urinary Tract Infections: Voided midstream urine will normally contain anaerobes from the urethral flora; therefore, collection of urine by percutaneous suprapubic bladder aspiration (shown in Figure 53) is necessary to accurately establish a diagnosis.

Female Genital Tract Infections: It is difficult or impossible to avoid normal flora in obtaining material from the endometrial cavity, and cultures are impossible to interpret in postpartum patients. Blood cultures may be helpful, as may culdocentesis cultures. In pyosalpinx or tubo-ovarian abscess, only cultures obtained during laparoscopy or other surgical procedures are reliable.

Techniques for transporting, processing, and identifying anaerobic bacteria are discussed in the next sections. Susceptibility testing of anaerobes is also discussed.

TRANSPORT OF ANAEROBIC SPECIMENS
Several methods are available (some commercially – appendix C) for transporting anaerobic specimens. Nevertheless, specimen transport is one of the weakest links in the chain from specimen collection to organism identification. Because some anaerobes can tolerate oxygen and survive for a number of days if kept moist, every laboratory will have some degree of success, regardless of the transport system used. However, isolation of relatively nonfastidious anaerobes such as *B fragilis* and *C perfringens* can cause more demanding anaerobes to be overlooked by providing a false sense of security that the laboratory's technique is good since it is isolating anaerobes. Because many anaerobes responsible for infections do not tolerate exposure to oxygen well, special transport methods are needed to ensure survival of the most fastidious anaerobes, from the time of specimen collection to the start of the analysis. Various means (Figures 67-73) for transporting anaerobic specimens will be described.

The *Transport Tube* (Figure 67) is prepared either by thoroughly flushing it with carbon dioxide or by placing it in an anaerobic chamber and then, after 48 hours, closing it with the rubber diaphragm and cap, as shown. The cap is then covered with foil and the assembled tube is sterilized. The tube may be prepared containing a small amount of prereduced peptone yeast medium to facilitate handling of small specimens and to incorporate an oxidation-reduction indicator (resazurin). Specimens are collected by needle and syringe and placed in the tube by inserting the needle through the diaphragm, after first expelling all air from the syringe and the lumen of the needle.

The *Vial Transporter* (Figure 68), which can be obtained from BBL or GIBCO, is a diaphragm-stoppered bottle with an anaerobic atmosphere that contains prereduced agar with a redox indicator. Systems using vials with liquid are also commercially available and provide convenient methods for anaerobic transport.

The *Anaerobic BioBag Transport* (Figure 69), which can be obtained from Marion Scientific Corp, is a plastic bag that can be used to transport a tissue specimen, a specimen in a syringe, or a specimen on a swab under sterile conditions. When the bag is sealed and the tubes containing the indicator, gas generator, and catalyst are crushed, an anaerobic atmosphere is provided.

A *Syringe Transport* (Figure 70) can be used if other

Figure 67. Transport tube.

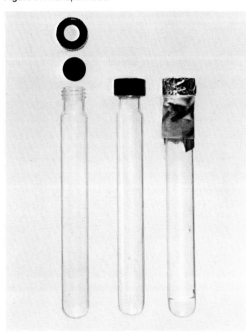

anaerobic transport systems are unavailable. Specimens can be taken to the laboratory in a syringe and needle. After the specimen is collected and all air is expelled from the syringe and needle, the needle is inserted into a sterile rubber stopper. However, specimens transported in this manner should be delivered to the laboratory *immediately* and set up in culture within 20 to 30 minutes.

The *Cotton Swab Transport Method* (Figure 71) requires a prereduced anaerobically sterilized (PRAS) swab prepared in one tube and a deep column of semi-solid PRAS Cary and Blair medium in another. After the specimen is collected on the anaerobic swab, it is immediately plunged gently into the Cary and Blair medium, and the tube is tightly stoppered and transported to the laboratory.

The *Vacutainer Anaerobic Transporter*, a double-tube system (Figure 72) that can be obtained from Becton-Dickinson and Co uses a cotton swab attached to a plunger in the stopper. When the stopper and swab are removed, the specimen can be collected with the swab, or tissue or fluid can be placed in the inner tube. The stopper is replaced, and the plunger is pushed down both to release the central tube — with the specimen — into the outer tube and to close the hole in the rubber stopper. The outer tube contains a mixture of H_2, CO_2, and N_2 and also palladium-covered catalyst pellets. The hydrogen combines with the oxygen introduced and forms water, thus achieving anaerobiosis.

The *Anaerobic, or GasPak, Jar* (Figure 73), which can be obtained from BBL, contains a GasPak envelope that generates H_2 and CO_2 after addition of water. A basket on the lid contains catalyst pellets that catalyze the conversion of H_2 and O_2 to water, thereby removing O_2 from the system. Vented lids are available that allow the evacuation-replacement system to be used to achieve anaerobiosis. The GasPak jar can also be used for transporting large specimens.

Figure 68. Vial transporter.

Figure 69. Anaerobic BioBag.

63

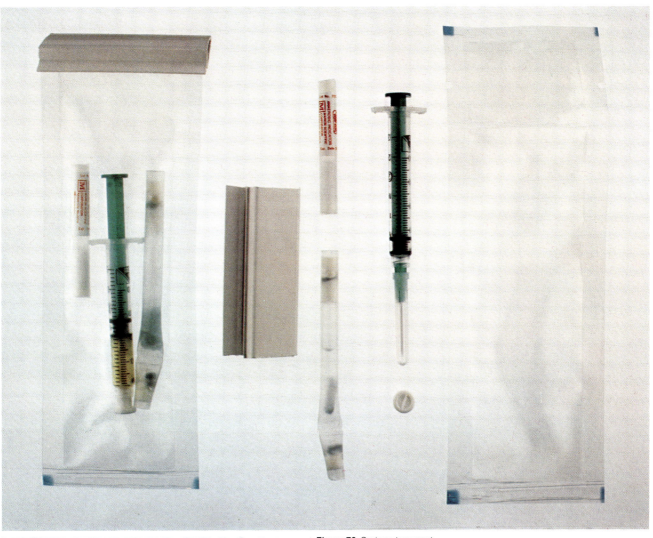

Figure 70. Syringe transport.

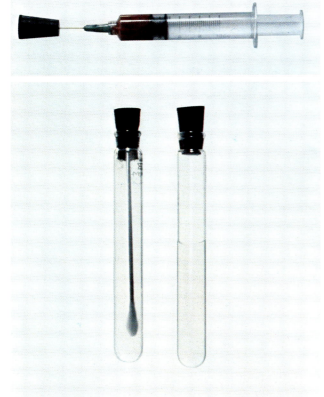

Figure 71. Cotton swab transport.

Figure 72. Vacutainer Anaerobic Transporter: a double-tube system.

Figure 72. Vacutainer Anaerobic Transporter: a double-tube system.

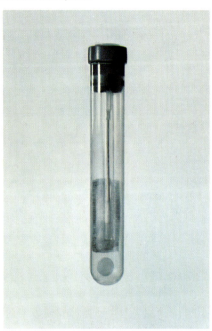

Finally, all specimens for anaerobic culture should be processed as soon as possible after collection because specimens frequently contain a mixed flora, including facultative organisms, most of which grow faster than anaerobes and may grow at room temperature.

SPECIMEN PROCESSING TECHNIQUES

The bibliography lists resources that give detailed procedures for handling anaerobes. A minimum processing scheme using commercially available media (appendix C) is shown in Table 18.

It is imperative that a Gram stain and a dark-field examination be done immediately on each specimen to make a preliminary identification of the infecting organisms. The results of these procedures are later compared with culture results. Dark-field examination is most useful because it shows cell morphology, motility, and forms that do not stain; yet it is the most neglected procedure in anaerobic bacteriology.

Good bench techniques such as surface plating and incubation in GasPak (Figure 73) or evacuation-replacement anaerobic jars are suitable for work-up of clinical specimens, because studies to date indicate that extremely oxygen-sensitive bacteria are seldom involved in human infections; *Clostridium difficile* may be an important exception. Accordingly, the use of the PRAS tube technique (Figure 74) or an anaerobic chamber (Figures 75 and 76), also referred to as a glove box, is not imperative unless *C difficile* is anticipated. It is advisable, however, to inoculate a tube of supplemented thioglycolate medium or a PRAS-enriched liquid medium, such as chopped-meat glucose. In the case of the PRAS tube, if inoculation by syringe and needle through the stopper of the sealed oxygen-free tube is not possible, an oxygen-free gassing apparatus must be used. These media are used as enrichment media so that if small numbers of bacteria are present but undetectable by plate inoculation, they may grow in the enriched

Figure 73. GasPak anaerobic jar.

broth and be detectable on subsequent plating. The broth tube also serves as a backup if oxygen enters an improperly sealed jar.

IDENTIFICATION OF ANAEROBIC BACTERIA

After colonial growth is obtained, the isolates are tested for purity and then subcultured to determine whether they are obligate anaerobes. Presumptive and definitive identification of anaerobes will be discussed after a brief discussion of other organisms that may be present.

Along with anaerobes, facultative organisms (those that grow well both in the presence and in the absence of air) are often present in human infections. These facultative organisms must also be isolated in pure culture and identified using appropriate methods. Some bacteria considered to be anaerobic can grow microaerophilically (under reduced oxygen tension) or are aerotolerant (can grow feebly aerobically) after initial anaerobic isolation. Examples of anaerobes that grow under microaerophilic conditions are *Actinomyces naeslundii, Arachnia propionica,* and some gram-positive cocci. An example of an aerotolerant anaerobe is *Clostridium histolyticum.*

Presumptive identification of anaerobes is often made on the basis of colonial morphology, cellular morphology, reaction to Gram stain, and other characteristics (appendices A and B). Definitive identification is also based on a number of morphologic and physiologic characteristics. Tests for pathogenicity and toxin production are necessary for definitive identification of certain clostridia. (These tests are available through reference laboratories.)

Identification of *Clostridium* depends on the demonstration of spores; this can be very difficult with certain of the species, such as *C perfringens* and *C ramosum.* Again, information provided in appendices A and B may be helpful in making a presumptive identification of a number of species of *Clostridium* that are of significance

---

Table 18.
Minimum processing of specimens for anaerobic culture by jar techniques

1. Direct examination of specimen
   (Gram stain and dark-field examination)

2. Culture

a Liquid enrichment	b Solid nonselective medium	c Solid selective medium
Supplemented thioglycolate medium or *PRAS chopped-meat glucose*	*Blood agar (brucella base)*	*Kanamycin-vancomycin laked blood agar* *Bacteroides bile-esculin agar*

Figure 74. Technique for inoculation of prereduced anaerobically sterilized (PRAS) tubes.

Figure 76. Diagram of principal parts of anaerobic chamber.

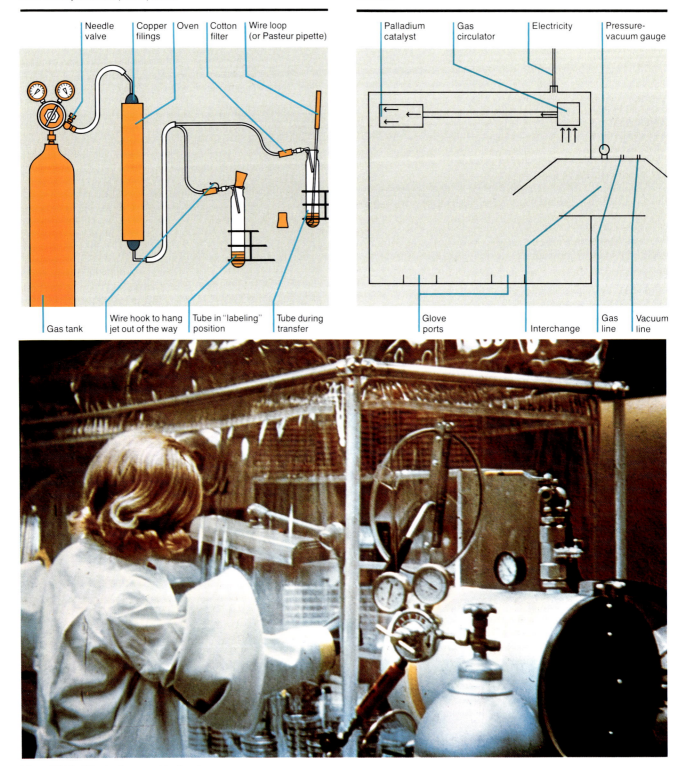

Needle valve | Copper filings | Oven | Cotton filter | Wire loop (or Pasteur pipette)

Gas tank | Wire hook to hang jet out of the way | Tube in "labeling" position | Tube during transfer

Palladium catalyst | Gas circulator | Electricity | Pressure-vacuum gauge

Glove ports | Interchange | Gas line | Vacuum line

Figure 75. Flexible plastic anaerobic chamber (or glove box).

in human disease. Careful definitive identification of *C botulinum* and *C tetani* depends on the demonstration of specific toxins. Definitive identification of other species of *Clostridium* depends on the demonstration of toxins; characteristics of pathogenicity, proteolytic activity, fermentation of various carbohydrates; other physiologic characteristics; and end products of metabolism.

Identification of nonsporeforming gram-positive bacilli is also difficult. As methods for classifying microorganisms have evolved, presumptive identification based on colonial and microscopic morphology is now known to be misleading. For example, bacteria that have been implicated in actinomycosis and other infections *(Actinomyces israelii, A naeslundii, Arachnia propionica,* and *Bifidobacterium eriksonii)* all have been classified as *Actinomyces* at some time in the past. However, by current taxonomic criteria (which include analysis of end products of metabolism), *Arachnia propionica* and *Bifidobacterium eriksonii* are no longer classified as *Actinomyces*. Current criteria for classification of genera of nonsporeforming gram-positive bacilli are based to a great extent on the major acid products of glucose metabolism (Table 19). Propionibacteria are often reported as "anaerobic diphtheroids." Frequently, they merely represent contamination of the specimen by anaerobes from skin; however, in some instances these organisms may be the causative agents in, or at least contributors to, infectious processes (particularly where there is an implanted prosthesis).

Generic classification of the nonsporeforming gram-negative bacilli also depends to some extent on the acid products of glucose metabolism. For example, *Fusobacterium* species produce large amounts of butyric acid, whereas most *Bacteroides* species do not. *Bacteroides putredinis, B splanchnicus,* and *B asaccharolyticus* produce small amounts of butyric acid but always with isobutyric and isovaleric acids. Identification of species and subspecies among these gram-negative

bacilli is based on a variety of characteristics such as growth in the presence of either bile or desoxycholate, susceptibility to antibiotics, fermentation of carbohydrates, and other physiologic reactions.

Anaerobic gram-positive cocci are divided into two genera. Cellular morphology and arrangement of cells are not adequate for generic identification. Speciation is based on a variety of physiologic characteristics such as fermentation of carbohydrates, indole production, nitrate reduction, and end products of glucose metabolism. A group of cocci frequently reported as "micro-aerophilic streptococci" are often implicated in serious infections such as endocarditis, brain abscess, and liver abscess. These cocci produce lactic acid as their major metabolic end product; they belong in the genus *Streptococcus*.

The most common anaerobic gram-negative coccus isolated from human infection is *Veillonella parvula,* which does not ferment hexoses but does convert lactate to propionate and pyruvate to acetate and propionate. Other anaerobic gram-negative cocci that are found less frequently are *Acidaminococcus fermentans* and *Megasphaera elsdenii.*

SUSCEPTIBILITY TESTING OF ANAEROBES

Anaerobic bacteria can no longer be considered to have predictable patterns of susceptibility to antimicrobial agents. An appreciable number of recent isolates of some of the more commonly encountered anaerobes, such as those in the *Bacteroides fragilis* group, *B melaninogenicus,* and peptococci, are now resistant to drugs that were once the drugs of choice. Although susceptibility testing is not usually feasible as a routine procedure for all anaerobic isolates, it is advisable when **1** isolates belong to the *B fragilis* group; **2** the infection is serious and life-threatening—such as endocarditis or a brain abscess; **3** prolonged therapy is anticipated; or **4** therapy with other drugs has failed.

Table 19.
Nonsporeforming gram-positive bacilli:
Major fatty and organic acids produced from glucose

Genus	Fatty and organic acids
Actinomyces	acetic, lactic, succinic
Arachnia	acetic, propionic
Bifidobacterium	acetic, lactic
Lactobacillus	lactic
Propionibacterium	acetic, propionic
Eubacterium	other than above: a. butyric, plus others b. acetic, formic c. no major acids

Susceptibility tests for anaerobic bacteria may be done by either broth or agar dilution techniques; micro broth trays with several dilutions of several drugs are available commercially. Disc diffusion tests similar to those being used for aerobic and facultative bacteria have been developed, but they require different setups for rapid and for slow-growing anaerobes. The simple broth-disc test involves regular antibiotic susceptibility test discs that are dropped into broth to achieve specific critical concentrations of drugs in one-tube tests; an organism's susceptibility or resistance can then be determined. These tests are easy to perform and appear to be reliable.

Therapy of Anaerobic Infections

Susceptibilities of commonly encountered anaerobes to some antimicrobial agents used to treat anaerobic infections are listed in Table 20. This table, based on in vitro determinations, shows the comparative susceptibilities of certain anaerobes to various drugs.

Penicillin G is the drug of choice for all anaerobic infections caused by susceptible microorganisms, although the *Bacteroides fragilis* group is most often resistant, and other gram-negative anaerobic bacilli may be resistant. On occasion, penicillin G has proved efficacious in *B fragilis* infections, but it is not recommended. Patients on penicillin therapy may, rarely, suffer serious hypersensitivity reactions (including anaphylaxis) that can be fatal.

Other penicillins and cephalosporins are not always as active as penicillin G (Figures 77 and 78); ampicillin and penicillin V are the only penicillins comparable to penicillin G in activity. The high blood levels achieved with carbenicillin, ticarcillin, mezlocillin, and piperacillin make them effective against 95% of *B fragilis* strains. Cefoxitin, a compound resistant to beta lactamases, is active against 90% to 95% of strains of *B fragilis*. Cefamandole is active against most anaerobes except those in the *B fragilis* group. Moxalactam, cefoperazone, cefotaxime, and other new beta-lactam agents are less active than cefoxitin against *B fragilis*; the exception is thienamycin, which has outstanding in vitro activity against anaerobes.

The clinical effectiveness of chloramphenicol in treating serious *B fragilis* infections and other serious anaerobic infections is well established, but serious toxicity may (rarely) accompany its use. For this reason, it is not listed as the drug of choice in Table 20.

Clindamycin is active against virtually all anaerobic gram-negative bacilli such as *Bacteroides* and *Fusobacterium* species. Clindamycin also shows good activity against anaerobic gram-positive nonsporeforming bacilli such as *Actinomyces*, *Propionibacterium*, and

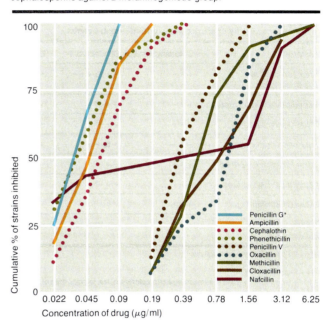

Figure 77. Activity of various penicillins and cephalosporins against *B melaninogenicus* group.

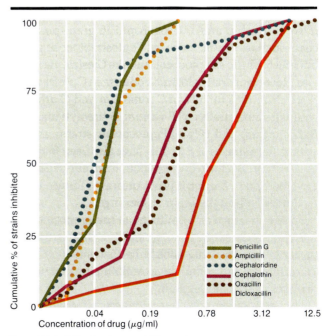

Figure 78. Activity of various penicillins and cephalosporins against anaerobic cocci.

*Recently, some strains have shown relative resistance to penicillin with MICs up to 25μg/ml.

Table 20.
In vitro susceptibility of anaerobes to antimicrobial agents[a]

Bacteria	Chloramphenicol	Clindamycin	Erythromycin✝	Metronidazole	Penicillin G	Tetracycline	Vancomycin✝
Microaerophilic & anaerobic cocci	+ + +	+ + to + + +	+ + to + + +	+ +	+ + + to + + + +	+ to + +	+ + +
Bacteroides fragilis group	+ + +	+ + +[b]	+ to + +	+ + +	+	+ to + +	+
Other *Bacteroides* sp	+ + +	+ + +[c]	+ + to + + +	+ + +	+ + to + + +[c]	+ +	+
Fusobacterium varium	+ + +	+ to + +	+	+ + +	+ + +[c]	+ +	+
Other *Fusobacterium* sp	+ + +	+ + +	+	+ + +	+ + + +	+ + +	+
Clostridium perfringens	+ + +	+ + +[b]	+ + +	+ + +	+ + + +[c]	+ +	+ + +
Other *Clostridium* sp	+ + +	+ +	+ + to + + +	+ + +	+ + +	+ +	+ + to + + +
Actinomyces and *Eubacterium*	+ + +	+ + to + + +	+ + +	+ to + +	+ + + +	+ + to + + +	+ + to + + +

a   + Poor or inconsistent activity
    + + Moderate activity
    + + + Good activity
    + + + + Good activity, good pharmacologic characteristics, low toxicity, drug of choice
b Rare strains are resistant
c A few strains are resistant
✝ Not approved by FDA for anaerobic infections

Eubacterium species, and it also shows activity against most anaerobic and microaerophilic gram-positive cocci such as *Peptococcus* sp, *Peptostreptococcus* sp, microaerophilic streptococci, and the *Clostridium* species. However, certain clostridia other than *C perfringens* (including 10% to 20% of *C ramosum* and *C difficile* strains); some strains of *F varium*, *B ureolyticus*, and *Eubacterium*; and 10% of *Peptococcus* strains are resistant. Rare strains of *B fragilis* have also been found to be resistant. Because clindamycin does not penetrate the blood-brain barrier, it is not indicated for use in infections affecting the brain or meninges.

Severe diarrhea and colitis have been reported following clindamycin therapy, and if either develops, therapy with the drug should be discontinued. (See Warning box and complete prescribing information at the back of this monograph.)

The pseudomembranous colitis associated with clindamycin or other antimicrobial therapy is caused primarily by *C difficile*. Oral vancomycin is effective in the treatment of pseudomembranous colitis, but there is a 20% relapse rate.

Metronidazole is active against all anaerobic bacteria except for 2% to 3% of anaerobic cocci and most strains of microaerophilic cocci, *Actinomyces*, *Arachnia*, *Propionibacterium*, and *Eubacterium*. It is the only agent that shows consistently good bactericidal activity against *B fragilis*. The drug is clinically effective in anaerobic infections, including those involving the central nervous system.

Tetracycline, once quite widely used in anaerobic infections, is now relegated to a much lower position because of the development of resistant strains of anaerobes. Tetracycline is presently indicated only for infections caused by strains known to be susceptible to it or in mild infections for which a therapeutic trial would be reasonable. Presently at our institution, half the strains of *B fragilis* are resistant to tetracycline. Some clostridia, *F varium*, anaerobic cocci (particularly the microaerophilic cocci), and *Eubacterium* are also resistant. Doxycycline and minocycline are more active than other tetracyclines, but susceptibility testing is still indicated to ensure activity.

The aminoglycosides are generally inactive against the majority of anaerobes.

Many infections in which anaerobes are found also involve facultative and aerobic microorganisms, and the nature of the nonanaerobic organisms will influence the choice of antimicrobial therapy. In general, antimicrobial therapy for anaerobic infections requires high-dose, prolonged treatment because of tissue necrosis and the tendency for relapse. Use of any antimicrobial agent may result in overgrowth of nonsusceptible organisms, including fungi.

Surgical therapy is extremely important. Abscesses must be drained and necrotic tissue debrided. Obstructions must be relieved and drainage established. Because of a marked tendency for abscess formation, so characteristic of anaerobic infection, repeated drainage procedures may be necessary.

Hyperbaric oxygen therapy may be useful in selected cases of gas gangrene when adequate debridement of necrotic tissue is not feasible. Hydrogen peroxide or zinc peroxide applied locally may effect dramatic improvement in certain difficult-to-manage superficial anaerobic infections.

Characteristics of some anaerobes involved in human disease

Organism	Synonyms	Colonial morphology and reaction on blood agar	Microscopic morphology	Other distinctive features
SPOREFORMING BACILLI				
Clostridium bifermentans	Bacillus bifermentans sporogenes	slightly raised, grey, irregular spreading edge, with narrow zone of hemolysis	gram-positive bacilli, with oval, subterminal spores	lecithinase produced on egg yolk agar, motile
Clostridium botulinum	Bacillus botulinus	colony morphology variable, depending on type; most strains hemolytic	large gram-positive bacillus, with oval subterminal spores	motile, lipase produced on egg yolk agar. Identification based on production of specific types of toxins
Clostridium difficile	Bacillus difficilis	slightly raised, white opaque, irregular shape, not hemolytic	slender, long, gram-positive bacilli, with parallel sides and oval terminal spores	negative for lecithinase and lipase; chartreuse fluorescence under ultraviolet light; cause of antimicrobial agent-associated pseudo-membranous colitis
Clostridium histolyticum	Bacillus histolyticus	a. smooth colonies: small transparent, glistening with irregular shape, and narrow zone of hemolysis b. rough colonies: raised, greyish center, flat edge, with rhizoids	gram-positive pleomorphic bacilli, with oval subterminal spores	feeble growth aerobically
Clostridium novyi	Clostridium oedematiens Bacillus oedematiens	irregular, round colonies, grey translucent, irregular surface, hemolytic, colonies of many strains are motile	gram-positive bacilli, with oval subterminal spores	freshly prepared media needed for isolation of type B
Clostridium perfringens	Clostridium welchii Welchia perfringens	low convex, slightly opaque, grey, shiny with entire to irregular edge, and wide double zone of hemolysis	large gram-positive bacilli, with blunt ends, spores (oval subterminal) rarely seen	lecithinase produced on egg yolk agar, not motile
Clostridium ramosum	Bacteroides terebrans Catenabacterium filamentosum Eubacterium filamentosum	convex, white, glistening, with entire edge, not hemolytic	pleomorphic gram-positive bacilli in chains, with swellings in cells	spores difficult to demonstrate
Clostridium septicum	Bacillus oedematis maligni Clostridium oedematis maligni Bacillus septicus	slightly raised, greyish, glistening, with irregular, rhizoid margins, hemolytic	gram-positive bacilli with small oval subterminal spores, long thin cells and lemon-shaped forms may be found	no distinctive features; identity based on series of morphologic and physiologic characteristics
Clostridium sordellii	Bacillus oedematis sporogenes Clostridium oedematis	slightly raised, grey, with irregular surface and spreading irregular edge, narrow zone of hemolysis	gram-positive bacilli, with oval subterminal spores	lecithinase produced on egg yolk agar, motile, may be distinguished from Clostridium bifermentans by its production of urease
Clostridium sporogenes	Bacillus sporogenes	raised, grey-yellow center, flat periphery, with rhizoid edge, variable hemolysis	gram-positive bacilli, with oval subterminal spores, filamentous forms in old cultures	lipase produced on egg yolk agar, identity based on series of morphologic and physiologic characteristics
Clostridium tetani		slightly raised, translucent, glistening, grey, with irregular edge, narrow zone of hemolysis	slender gram-positive bacilli of varying length, with round terminal spores	many strains swarm over the surface of agar

*facultative or microaerophilic

(continued)

Characteristics of some anaerobes involved in human disease

Organism	Synonyms	Colonial morphology and reaction on blood agar	Microscopic morphology	Other distinctive features
NONSPOREFORMING BACILLI				
Actinomyces israelii	Streptothrix israelii	convex, rough, "molar tooth" colonies	long gram-positive bacilli, some with branching	colonies on brain heart infusion agar: day 2 (microscopic) filamentous "spider" colonies; day 7 (macroscopic) colonies heaped, lobate, "molar tooth"
Actinomyces naeslundii*		smooth–low convex, white transparent, with entire edge, no hemolysis; rough–raised, irregular, no hemolysis	moderately long gram-positive bacilli, with many short branches	colonies on brain heart infusion agar: day 2 (microscopic) filamentous "spider" colonies, with dense center; day 7 (macroscopic) smooth and convex
Arachnia propionica*	Actinomyces propionicus		gram-positive bacilli, with varying degrees of branching; club-shaped bacilli common	colonies on brain heart infusion agar: day 2 (microscopic) filamentous "spider" colonies; day 7 (macroscopic) may form either smooth, convex, or heaped or "molar tooth" colonies
Bifidobacterium adolescentis		slightly convex, white, smooth, with entire edge, not hemolytic	gram-positive bacilli, with bifurcated ends	
Bifidobacterium eriksonii	Actinomyces eriksonii	convex, white, glistening, irregular edge, may be slightly hemolytic	gram-positive bacilli, with clubbed or bifurcated ends	colonies on brain heart infusion agar: day 2 (microscopic) flat, granular, with dense center, irregular edge; day 7 (macroscopic) smooth and convex
Eubacterium alactolyticum	Ramibacterium alactolyticum	convex, opaque, glistening, entire edge	gram-positive bacilli, pleomorphic, with cells in V-arrangements	
Eubacterium lentum	Bacteroides lentus	raised to low convex, slightly opaque, smooth, with slightly irregular edge, not hemolytic	gram-positive, slightly pleomorphic coccobacilli, in chains	
Eubacterium limosum	Bacteroides limosus Butyribacterium rettgeri	convex, translucent to white, entire edge, not hemolytic	gram-positive bacilli, pleomorphic, in pairs and short chains	
Lactobacillus catenaforme	Catenabacterium catenaforme	convex, slightly translucent, with entire edge	gram-positive bacilli, pleomorphic, often in long chains	
Propionibacterium acnes	Corynebacterium acnes, "anaerobic diphtheroid"	convex, white to pink, shiny, opaque, entire edge, may be hemolytic	gram-positive bacilli, pleomorphic, club-shaped	
Bacteroides fragilis group	Ristella fragilis Eggerthella convexus Fusiformis fragilis	convex, white to grey, translucent, glistening, may be slightly hemolytic	gram-negative bacilli, rounded ends, may be pleomorphic	usually resistant to penicillin, growth often stimulated by bile
Bacteroides melaninogenicus-asaccharolyticus group	Bacteroides nigrescens Fusiformis melaninogenicus	convex, brown to black after 5-7 days, may be glistening or dull, hemolytic	gram-negative bacilli, rounded ends, often coccobacillary	young colonies exhibit brick-red fluorescence under ultra-violet light (Wood's lamp), growth inhibited by bile
Bacteroides oralis		convex, yellowish to translucent, smooth, glistening, may be hemolytic	gram-negative bacilli, some coccobacillary, some elongated	iodophilic polysaccharide produced from glucose, growth inhibited by bile

Characteristics of some anaerobes involved in human disease

Organism	Synonyms	Colonial morphology and reaction on blood agar	Microscopic morphology	Other distinctive features
Bacteroides pneumosintes	*Dialister pneumosintes*	pinpoint, convex, shiny, transparent, no hemolysis, hand-lens or stereoscopic microscope necessary for detection	tiny gram-negative bacilli, negative staining often required for detection	filterable through Berkfeld V and N filters
Bacteroides putredinis	*Bacillus putredinis* *Ristella putredinis*	pinpoint, low convex, grey, translucent, dull, with slightly irregular edges	gram-negative bacilli, pleomorphic	sensitive to penicillin, growth not inhibited by bile
Bacteroides ureolyticus	*Bacteroides corrodens*	pinpoint with edges spreading and eroding into agar, not hemolytic	gram-negative bacilli, rounded ends	smooth colony variants appear frequently on subculture
Fusobacterium mortiferum	*Sphaerophorus mortiferus* *Sphaerophorus freundii*	convex, slightly opaque center, with translucent spreading, irregular edge, "fried egg" appearance, no hemolysis	gram-negative bacilli, highly pleomorphic, with bizarre forms and round bodies	most strains sensitive to penicillin, growth in bile and in desoxycholate
Fusobacterium necrophorum	*Sphaerophorus necrophorus* *Sphaerophorus (Bacteroides) funduliformis* *Necrobacterium (Bacillus) funduliformis* *Fusiformis necrophorus*	convex, umbonate, with raised opaque center, translucent, entire edge, hemolytic or alpha-hemolytic	gram-negative bacilli, with rounded to tapered ends, pleomorphic, with round bodies	lipase produced on egg yolk agar by most strains
Fusobacterium nucleatum	*Fusiformis fusiformis* "fusiform bacillus" *Fusobacterium fusiforme*	convex, glistening, with internal iridescent flecking or raised, opaque "bread crumb" colonies, alpha-hemolytic	gram-negative bacilli, with tapered ends	sensitive to penicillin, growth inhibited by bile
Fusobacterium varium	*Sphaerophorus varius* *Bacteroides varius*	flat to low convex, slightly opaque center, translucent, spreading irregular edge, "fried egg" appearance, occasional strains slightly hemolytic or alpha-hemolytic	gram-negative bacilli, highly pleomorphic, with bizarre forms and round bodies	most strains sensitive to penicillin, growth in bile and in desoxycholate
ANAEROBIC COCCI *Peptococcus species*	*Micrococcus* species *Staphylococcus* species (eg, *S anaerobius*)	convex, grey to white, shiny with entire edges, may be hemolytic or alpha-hemolytic, opaque	small to large gram-positive cocci, singles, pairs, irregular clumps	species identity based on series of physiological characteristics
Peptostreptococcus species	*Streptococcus* species *Diplococcus* species *Micrococcus* species (eg, *M foetidus*)	convex, grey to white, shiny or dull, with entire edges, may be hemolytic or alpha-hemolytic, opaque	small to large gram-positive cocci, pairs and chains	species identity based on series of physiological characteristics
Veillonella parvula	*Neisseria* species *Micrococcus* species *Staphylococcus* species (eg, *S parvulus*)	convex, translucent, glistening, with entire edges, not hemolytic	small gram-negative cocci, pairs, short chains, and irregular clumps	growth stimulated by lactate or pyruvate

*facultative or microaerophilic

**Illustrated colonial and Gram-stain morphology
of some anaerobes involved in human disease**

Colonial morphology

Figure 79.   *Clostridium sordellii*

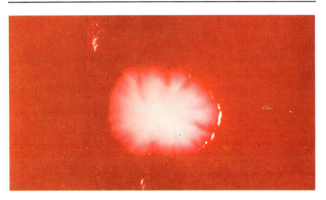

Figure 80.   *Clostridium bifermentans*

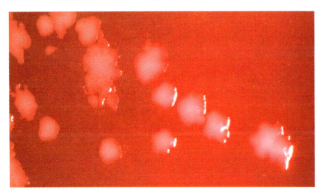

Figure 81.   *Clostridium difficile*

Figure 82.   *Actinomyces israelii*

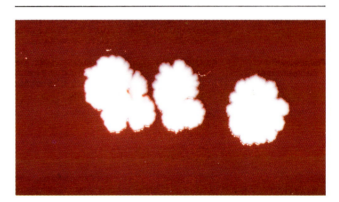

Figure 83.   *Propionibacterium acnes*

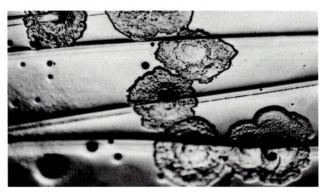

Figure 84.   *Bacteroides ureolyticus* – showing etched growth (rough)

77

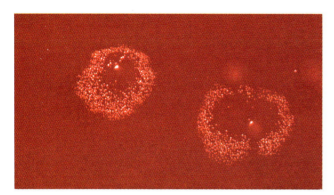

Figure 85. *Bacteroides ureolyticus*

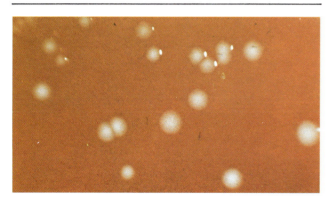

Figure 86. *Bacteroides fragilis*

Figure 87. *Bacteroides melaninogenicus*

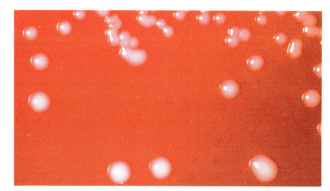

Figure 88. *Bacteroides ruminicola* ss *brevis*

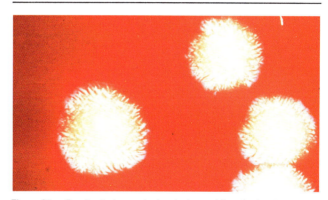

Figure 89. *Fusobacterium nucleatum* (note speckling of colony)

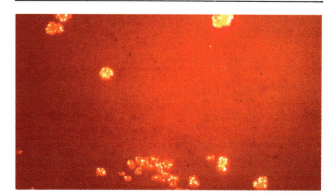

Figure 90. *Fusobacterium nucleatum* ("bread crumb" colony)

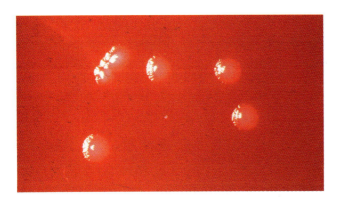

Figure 91. *Fusobacterium necrophorum*

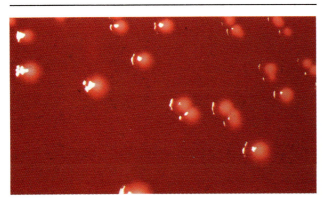

Figure 92. *Fusobacterium varium*

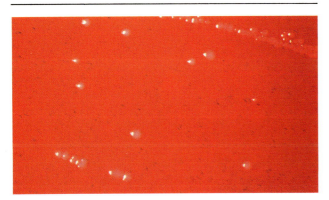

Figure 93. *Peptococcus asaccharolyticus*

Figure 94. *Peptostreptococcus anaerobius*

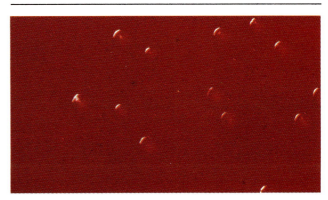

Figure 95. *Veillonella parvula*

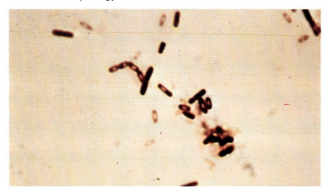

Figure 96. *Clostridium bifermentans.* Note destained, gram-negative forms

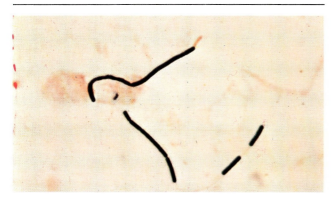

Figure 97. *Clostridium botulinum*

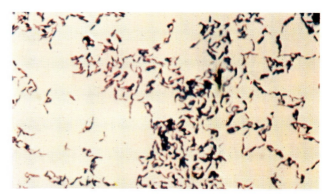

Figure 100. *Propionibacterium acnes*

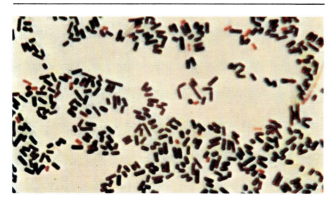

Figure 98. *Clostridium perfringens.* Note destained, gram-negative forms

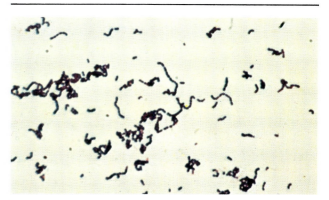

Figure 101. *Peptostreptococcus anaerobius*

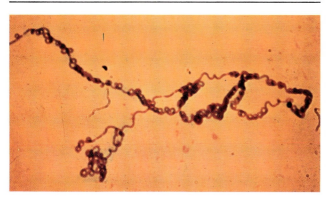

Figure 99. *Eubacterium species*

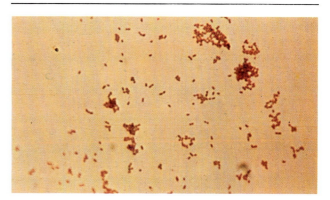

Figure 102. *Bacteroides melaninogenicus* (from blood agar plate)

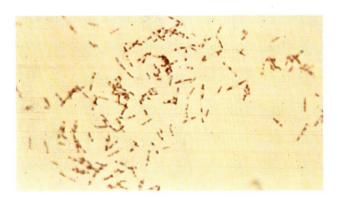

Figure 103. *Bacteroides fragilis* (from thioglycolate medium)

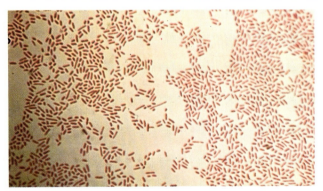

Figure 104. *Bacteroides fragilis* (from blood agar plate)

Figure 105. *Fusobacterium nucleatum*

Figure 106. *Fusobacterium nucleatum*

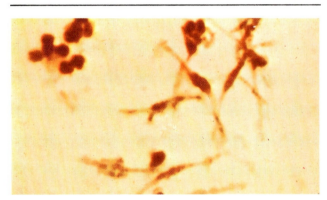

Figure 107. *Fusobacterium mortiferum*

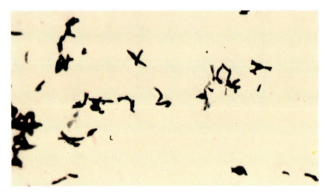

Figure 108. *Bifidobacterium* sp

**Sources of commercially available
anaerobic equipment and media**

Anaerobic Equipment

Bellco Glass, Inc (roll-tube equipment)
340 Edrudo Road
Vineland, NJ 08360

BBL Microbiology Systems (anaerobic jars and GasPaks)
Division of Becton, Dickinson and Co
PO Box 243
Cockeysville, MD 21030

Capco Instruments (anaerobic chambers)
PO Box 9093
Sunnyvale, CA 94086

Coy Manufacturing (anaerobic chambers)
1393 Harpst Street
Ann Arbor, MI 48104

Forma Scientific, Inc (anaerobic chambers)
Division of Mallinckrodt, Inc
PO Box 649
Marietta, OH 45750

Kontes Company (mobile anaerobe laboratory)
Spruce Street
Vineland, NJ 08360

Dehydrated Media

BBL Microbiology Systems
PO Box 243
Cockeysville, MD 21030

Difco Laboratories
PO Box 1058A
Detroit, MI 48232

GIBCO Diagnostics
2801 Industrial Drive
Madison, WI 53713

Nissui Seiyaku Co, Ltd
5-11 Komagome, 2 Chome
Toshima-Ku
Tokyo 170, Japan

Prepared Anaerobic Media and Transport Systems

Anaerobe Systems
3074A Scott Boulevard
Santa Clara, CA 95050

Becton, Dickinson and Co
Rutherford, NJ 07070

Carr-Scarborough Microbiologicals, Inc
PO Box 581
Forest Park, GA 30050

GIBCO Diagnostics
2801 Industrial Drive
Madison, WI 53713

Marion Scientific Co
Marion Laboratories, Inc
10236 Bunker Ridge Road
Kansas City, MO 64137

Nolan Biological Laboratories, Inc
8 La Vista Perimeter Office Park
Tucker, GA 30084

Scott Laboratories, Inc
771 Main Street
Fiskeville, RI 02823

Acknowledgments

Figures 1 and 3. Finegold SM: Intracranial abscess, in Hoeprich PD (ed): *Infectious Diseases,* Hagerstown, MD, Harper & Row Publishers Inc, 1972, pp 965-969.

Figures 2, 3, 4, and 5. Neurosurgery Departments, UCLA Medical Center and Wadsworth VA Hospital, Los Angeles, California.

Figures 10, 11, 12, and 14. James B. Taylor, Consultant,Long Beach Veterans Administration Hospital, Long Beach, California.

Table 12. Modified from MacLennan JD; *Bacteriol Rev* 26:177-276, 1962.

Figure 13. Low RC, Dodds TC: *Atlas of Bacteriology.* Edinburgh, E & S Livingston Ltd, 1947.

Figure 16. John G. Bartlett, Johns Hopkins University School of Medicine, Baltimore, MD.

Figures 28A, 28B, 28C, 32, 42, 54, 59, 62, and 66. Martin C. McHenry, Cleveland Clinic, Cleveland, OH.

Figures 29, 63B, 77, and 107. Finegold SM: Infections due to anaerobes. *Med Times* 96:174-187, 1968.

Figures 38A, 38B, and 41. Patterson DK, Ozeran RS, Glantz GJ, Miller AB, Finegold SM: Pyogenic liver abscess due to microaerophilic streptococci. *Ann Surg* 165:362-376, 1967.

Figures 44 and 45. Finegold SM: Appendicitis and diverticulitis, in Hoeprich PD (ed): *Infectious Diseases,* Hagerstown, MD, Harper & Row Publishers Inc, 1972, pp 693-698.

Figure 46. Finegold SM, Dineen P: Subphrenic and other intra-abdominal abscesses, in Hoeprich PD (ed): *Infectious Diseases,* Hagerstown, MD, Harper & Row Publishers Inc, 1972, pp 703-708.

Figures 50, 65, 79, 96, 97, 102, 103, and 105. Center for Disease Control, Atlanta, GA.

Figure 52. Polter DE, Boyle JD, Miller LG, Finegold SM: Anaerobic bacteria as cause of the blind loop syndrome. *Gastroenterology* 54:1148-1154, 1968.

Figure 55. Altemeier WA, Fullen WD: Prevention and treatment of gas gangrene. *JAMA* 217:806-813, 1971.

Figure 74. Finegold SM, Sutter VL, Attebery HR, et al: Isolation of anaerobic bacteria, in Lennette EH, Spaulding EH, Truant JP (eds): *Manual of Cinical Microbiology.* Washington, American Society for Microbiology, 1970, chap. 32, pp 365-375.

Bibliography

Ajello L, Georg LK, Kaplan W, et al: Mycotic infections (re: Actinomyces), in Bodily HL, Updyke EL, Mason JO (eds): *Diagnostic Procedures for Bacterial, Mycotic and Parasitic Infections*, ed 5. New York, Am Publ Health Assoc, 1970.

Balows A, DeHaan RM, Dowell VR Jr, et al: *Anaerobic Bacteria: Role in Disease*. Springfield, Ill, Charles C Thomas, 1974.

Bartlett JG, Finegold SM: Anaerobic pleuropulmonary infections. *Medicine* 51:413-450, 1972.

Bartlett JG, Gorbach SL: Anaerobic infections of the head and neck. *Otolaryngol Clin North Am* 9:655-678, 1976.

Bartlett JG, Gorbach SL, Finegold SM: The bacteriology of aspiration pneumonia. *Am J Med* 56:202-207, 1974.

Bartlett JG, Onderdonk AB, Cisneros RL, et al: Clindamycin-associated colitis due to a toxin-producing species of *Clostridium* in hamsters. *J Infect Dis* 136:701-705, 1977.

Beerens H, Tahon-Castel M: *Infections Humaines à Bactéries Anaérobies non Toxigènes*. Bruxelles, Presses Académiques Européennes, 1965.

Bornstein DL, Weinberg AN, Swartz MN, et al: Anaerobic infections – review of current experience. *Medicine* 43:207-232, 1964.

Burnett GW, Scherp HW, Schuster GS: *Oral Microbiology and Infectious Disease*, ed 4. Baltimore, Williams & Wilkins Co, 1976.

Chow AW, Malkasian KL, Marshall JR, et al: The bacteriology of acute pelvic inflammatory disease. *Am J Obstet Gynecol* 122:876-879, 1975.

Dowell VR Jr: Anaerobic infections, in Bodily HL, Updyke EL, Mason JO (eds): *Diagnostic Procedures for Bacterial, Mycotic and Parasitic Infections*, ed 5. New York, Am Publ Health Assoc, 1970.

Dowell VR Jr, Hawkins TM: *Laboratory Methods in Anaerobic Bacteriology*. CDC Laboratory Manual, No. 74-8272. US Department of Health, Education and Welfare, Public Health Service, 1974.

Felner JM, Dowell VR Jr: Anaerobic bacterial endocarditis. *N Engl J Med* 283:1188-1192, 1970.

Felner JM, Dowell VR Jr: *Bacteroides* bacteremia. *Am J Med* 50:787-796, 1971.

Finegold SM: *Anaerobic Bacteria in Human Disease*. New York, Academic Press, 1977.

Finegold SM: Gram-negative anaerobic rods – Bacteroidaceae, in Sonnenwirth AC, Jarrett L (eds): *Gradwohl's Clinical Laboratory Methods and Diagnosis*, ed 8. St. Louis, CV Mosby Co, 1980.

Finegold SM, Marsh VH, Bartlett JG: Anaerobic Infections in the Compromised Host. *Proc Internat Conf on Nosocomial Infections*. Chicago, Am Hosp Assoc, 1971, pp 123-134.

Finegold SM, Miller LG, Merrill SL, et al: Significance of anaerobic and capnophilic bacteria isolated from the urinary tract, in Kass EH (ed): *Progress in Pyelonephritis*, Philadelphia, FA Davis Co, 1965, pp 159-178.

Finegold SM, Sutter VL, Cato EP, et al: Anaerobic bacteria, in Graber CD (ed): *Rapid Diagnostic Methods In Medical Microbiology*, Baltimore, Williams & Wilkins Co. 1970.

Frederick J, Braude AI: Anaerobic infection of the paranasal sinuses. *N Engl J Med* 290:135-137, 1974.

George WL, Rolfe RD, Mulligan ME, et al: Infectious diseases 1979 – antimicrobial agent-induced colitis: An update. *J Infect Dis* 140:266-268, 1979.

George WL, Sutter VL, Finegold SM: Antimicrobial agent-induced diarrhea – a bacterial disease. *J Infect Dis* 136:822-828, 1977.

Goldsand G, Braude AI: Anaerobic infections. *Disease a Month*, November 1966.

Gorbach SL, Bartlett JG: Anaerobic infections. *N Engl J Med* 290:1177-1184, 1237-1245, 1289-1294, 1974.

Heineman HS, Braude AI: Anaerobic infection of the brain. *Am J Med* 35: 682-697, 1963.

Holdeman LV, Cato EP, Moore WEC: *Anaerobe Laboratory Manual*, ed 4. Blacksburg, Va, Anaerobe Laboratory, VPI and State Univ, 1977.

Kurzynski TA, Yrios JW, Helstad AG, et al: Aerobically incubated thioglycolate broth disk method for antibiotic susceptibility testing of anaerobes. *Antimicrob Agents Chemother* 10:727-732, 1976.

Lennette EH, Balows A, Hausler WJ Jr, Truant JP (eds): *Manual of Clinical Microbiology*, ed 3. Washington, American Society for Microbiology, 1980.

Lewis RP, Sutter VL, Finegold SM: Bone infections involving anaerobic bacteria. *Medicine* 57:279-305, 1978.

Lorber B, Swenson RM: The bacteriology of intra-abdominal infections. *Surg Clin North Am* 55:1349-1354, 1975.

MacLennan JD: The histotoxic clostridial infections of man. *Bacteriol Rev* 26:177-276, 1962.

Marcoux JA, Zabransky RJ, Washington JA II, et al: *Bacteroides* bacteremia: A review of 123 cases. *Minn Med* 53:1169-1176, 1970.

Martin WJ: Isolation and identification of anaerobic bacteria in the clinical laboratory, a two-year experience. *Mayo Clin Proc* 49:300-308, 1974.

Phillips I, Sussman M (eds): *Infection With Nonsporing Anaerobic Bacteria*, A symposium of the British Society for Antimicrobial Chemotherapy. Edinburgh, Churchill-Livingstone, 1974.

Rosebury T: *Microorganisms Indigenous to Man*. New York, McGraw-Hill Book Co, 1962.

Rotheram EB Jr, Schick SF: Nonclostridial anaerobic bacteria in septic abortion. *Am J Med* 46:80-89, 1969.

Sabbaj J, Sutter VL, Finegold SM: Anaerobic pyogenic liver abscess. *Ann Intern Med* 77:629-638, 1972.

Saksena DS, Block MA, McHenry MC, et al: Bacteroidaceae: Anaerobic organisms encountered in surgical infections. *Surgery* 63:261-267, 1968.

Shapton DA, Board RG (eds): *Isolation of Anaerobes.* New York, Academic Press Inc, 1971.

Smith LDS: *The Pathogenic Anaerobic Bacteria,* ed 2. Springfield, Ill, Charles C Thomas, 1975.

Socransky SS: Microbiology of periodontal disease – present status and future considerations. *J Periodontol* 48:497-504, 1977.

Sterne M, Batty I: *Pathogenic Clostridia.* London, Butterworths, 1975.

Stokes EJ: Anaerobes in routine diagnostic cultures. *Lancet* 1:668-670, 1958.

Sutter VL, Finegold SM: Susceptibility of anaerobic bacteria to 23 antimicrobial agents. *Antimicrob Agents Chemother* 10:736-752, 1976.

Sutter VL, Citron DM, Finegold SM: *Wadsworth Anaerobic Bacteriology Manual,* ed 3. St. Louis, CV Mosby Co, 1980.

Washington JA II: Bacteremia due to anaerobic, unusual and fastidious bacteria, in Sonnenwirth AC (ed): *Bacteremia, Laboratory and Clinical Aspects.* Springfield, Illinois, Charles C Thomas, 1973, pp 47-60.

Wilkins TD, Thiel T: Modified broth disk method for testing the antibiotic susceptibility of anaerobic bacteria. *Antimicrob Agents Chemother* 3:350-356, 1973.

Willis AT: *Anaerobic Bacteriology: Clinical and Laboratory Practice,* ed 3. London, Butterworths, 1977.

Willis AT: *Clostridia of Wound Infection.* London, Butterworths, 1969.

Ziment I, Davis A, Finegold SM: Joint infection by anaerobic bacteria: A case report and review of the literature. *Arthritis Rheum* 12:627-635, 1969.

Index

86